W9-CPO-896

The Challenge of New Technology

Also by David Simpson

General Equilibrium Analysis (Blackwell, 1975)
The Political Economy of Growth (Blackwell, 1983)

The Challenge of New Technology

David Simpson
Professor of Economics
University of Strathclyde

Jim Love
Fraser of Allander Institute
University of Strathclyde

Jim Walker
Fraser of Allander Institute
University of Strathclyde

WHEATSHEAF BOOKS

WESTVIEW PRESS
Boulder, Colorado

First published in Great Britain in 1987 by
WHEATSHEAF BOOKS LIMITED
A MEMBER OF THE HARVESTER PRESS PUBLISHING GROUP
Publisher: John Spiers
16 Ship Street, Brighton, Sussex
and in the United States of America by
WESTVIEW PRESS, INC.
5500 Central Avenue, Boulder, Colorado 80301
Frederick A. Praeger, Publisher

British Library Cataloguing in Publication Data

Simpson, David, *1936-*
The challenge of new technology.
1. Technological innovations — Social aspects
I. Title II. Love, J. III. Walker, Jim
303.4'83 T173.8

ISBN 0-7450-0344-3
ISBN 0-7450-0377-X Pbk

WESTVIEW PRESS
ISBN 0-8133-0613-2
LCN 87-50655

Typeset in 11/12 point Times by Witwell Ltd, Liverpool
Printed in Great Britain by Mackays of Chatham Ltd, Kent

Contents

Preface

There is a vast literature, whose rate of output shows no sign of diminishing, of forecasts of the consequences of new technology. These forecasts are for the most part written by scientists, technologists and engineers, and their distinguishing characteristic is a lack of uncertainty about the future. The role of the human factor in the implementation of the technology is also noticeable by its absence.

The present book differs from such literature in two important respects. It does not deal in speculations about the future; instead it looks at what has actually happened when new technology has been deployed in an industrial and commercial environment. Secondly, it is written by economists, who are perhaps less sanguine about the predictability of human behaviour than their colleagues in the natural sciences. The book starts from the assumption that the social and economic consequences of new technology—whether for good or for evil—will depend on decisions made by human beings, whether as individuals, or as members of such collective groups as firms, departments of government, households, trade unions and other organisations. The advent of new technology creates challenges for individuals and organisations in the shape of opportunities for gain and prospects of loss. In this book, we set out to identify some of the effects which have been evident hitherto, as well as some of the challenges.

The book is built on two foundations. The authors have spent more than three years surveying the effects of the deployment of new technology in industry and commerce. In the course of this survey, more than 240 interviews, each of approximately 40 minutes' duration, with managers, shopfloor workers, trade union representatives and office staff were carried out and recorded on tape. We conducted the interviews in eight firms in each of five industries: electronics, printing, engineering, consumer goods and financial services. The firms that we interviewed ranged in ownership type from state-owned through multinational to private unquoted companies, and in size from employing 14 to employing 8000 people. The interviews were carried out in 1984 and 1985. While all of these interviews were conducted in the United Kingdom, we are reasonably confident that most of our conclusions apply generally to the industrialised economies of the world, which are converging fast not only in their technology but also in their patterns of economic and social behaviour.

Secondly, we have tried to read as much of the worthwhile English-language literature on our subject as we were able, although no doubt much valuable material has escaped our attention. We have therefore borrowed unashamedly from other people's ideas. We should like to believe that the book represents a reasonably balanced presentation of the relevant facts and opinions seen critically from different points of view and informed by our own experience.

In terms of method we have followed the example of Rosabeth Kanter, who wrote: 'in making choices about what material to use to express and delineate my ideas, I lean towards rendering those dramas of life in the corporation which would make my conclusions come alive, which would cause readers to believe me, not because of my numbers but because of the echoes of my ideas in their own experience.'[1]

Although there is a list of references and suggestions for further reading at the end of this book, we cannot possibly acknowledge all the intellectual debts which we have incurred. There are, however, some particularly influential works, upon whose ideas we have drawn freely, and which we now acknowledge formally: Simon Nora, *The Computerisation of Society*; Peter Drucker, *Innovation and Entrepreneurship*; J. Northcott, M. Fogarty and M. Trevor, *Chips and Jobs*; N. Rosenberg, *Inside the Black Box*; P. Large, *The Micro Revolution Revisited*; E. Braun and S. Macdonald, *Revolution in Miniature*; and T. Forester (ed.), *The Information Technology Revolution.*

There are a number of individuals whose advice and encouragement we should also like to acknowledge. We should like to thank Sir Bruce Williams, Director of the Technical Change Centre, London, during the period of this study. We should also have liked to have mentioned the names of other individuals who were particularly helpful to us during the survey period. Unfortunately to name them would be to disclose the identity of some of the firms which took part, and we have made it a rule to respect the confidentiality of every interviewee. To them and to all those who helped us we can only express our heartfelt thanks for their time and trouble.

To two colleagues who took part in our project at an earlier stage, Robert Crawford and Frank Harrigan, we should also like to express our thanks. Robert Crawford made such a large contribution to this project that he would have been justified in being included as a co-author, a privilege which he has unselfishly relinquished. Another colleague, John Scouller, kindly allowed us to look at the manuscript of his forthcoming book on industrial policy. Needless to say, none of the people mentioned bears any responsibility for the opinions expressed in this book, or for any errors of fact. These remain the sole responsibility of the authors.

We should like to express our particular appreciation to the Trustees of the Leverhulme Trust, without whose financial support our project would not have been possible.

Finally, we should like to thank Mrs Jean Nelson for preparing and revising early drafts of this manuscript.

David Simpson
Jim Love
Jim Walker

The Fraser of Allander Institute
and Department of Economics,
University of Strathclyde.

July 1986

NOTES

1. R. M. Kanter (1983), pp. 384, 385.

PART I
Introduction

1 Introduction

All western countries have now embarked on a great revolution in technology. The main areas of rapid technical progress at the present time include microelectronics, information technology, biotechnology, materials technology and energy technology. At the same time, developments in these basic fields are being applied to other areas of dramatic progress such as space technology, and the technology of exploring and exploiting the sea bed.

Technological revolutions in the past have brought about far-reaching economic and social changes. There is every reason to believe that the present revolution in technology will have equally widespread consequences, with perhaps a more rapid momentum than any of its predecessors. Nowhere has the rate of progress been faster than in the microelectronics industry which has evolved in the last forty years. Almost ten years ago Robert Noyce, Chairman of Intel Corporation observed:

> Today's microcomputer ... has more computing capacity than the first large electronic computer. It is 20 times faster, has a larger memory, is thousands of times more reliable, consumes the power of a light bulb rather than a locomotive, occupies 1/30,000 the volume, and costs 1/10,000 as much.

It has been estimated that if the same rate of technical progress had been achieved in the motor car industry, then a Rolls Royce today would cost approximately 15 cents and would be able to achieve a fuel consumption performance of several thousand miles per gallon. Everyone who works in business is aware of the quickening pace of activity, where it seems that the more senior the position the more intense the pressures. Even the individual who is immune from the pressures of business life is aware of the increased bustle and pace of day-to-day living. Developments in technology have played an important part in bringing this about.

While the challenges posed by present technical developments are many, the important issues are not primarily technical, but economic and social. These issues are also the least well understood. The benefits of new technology depend not just on the technology itself but on if, when, how and how well it is implemented. Whether we are talking about a housewife buying a programmable washing-machine, a government deciding what type of telecommunications system to adopt for a whole country, or an engineer deciding what kinds of improvement he can make in the operation of a piece of new machinery that has just been installed in his workplace, the effects of new technology are very much what people make of them. Where new technology is deployed, the resulting gains in productivity do not depend simply on the physical characteristics of the new technology itself. Studies have shown that the productivity of identical equipment in different factories can vary greatly. As we shall argue in this book, the decisive factors include the attitude of workers and management, and the organisation of production. The productivity of any given piece of new technology also depends on such associated factors as the provision of specialist services for the financing, distribution and marketing of the product.

This book is about how people react to the opportunities, and sometimes the threats, which new technology poses for them. It is not about the implementation of the whole range of new technologies; there is simply not the space to do this. Throughout this book the phrase 'new technology' is used to designate microprocessor technology and the related information-intensification process.

A microprocessor consists of an integrated circuit that makes up the central processing unit of a microcomputer. It may be used directly for that purpose or it may be used to perform other functions. An integrated circuit is one in which the functions of several discrete components are performed within a single piece of semiconductor material, that is material that may either conduct or not conduct a current. A microprocessor is a tiny fragment (chip) of material, usually of silicon, which contains all the circuitry necessary to perform the processing functions of a computer. Other chips have been developed to act as data storage or memory devices, and still others as controllers of input/output functions. Taking them together one has a very small and very cheap computer. For example, the Hewlett-Packard 9000 microcomputer has as its central processing unit a single microprocessor chip. The HP 9000 costs around $40,000, sits on a desktop and can perform 1 million instructions per second, a capability which allows it to compete with larger computers that have dozens of chips in their central processing units, are the size of refrigerators and cost hundreds of thousands of dollars.

Although it is too soon to say if the microprocessor is the most important technological innovation of recent years, it is certainly the common factor which has speeded the development of many others. The whole range of microprocessor-controlled and related technologies is extraordinarily diverse and it is worthwhile looking very briefly at its history.

The microprocessor is not really a single innovation but is the result of a long series of linked innovations. There is a continuous chain of new electronic devices based on semiconductor materials which stretches from the 'cat's whisker' used in radio sets at the turn of this century, to the latest and most astonishing achievements of modern microelectronics. The totality of these innovations is so massive as to make most single technological innovations pale into insignificance by comparison. Braun and Macdonald (1978) have remarked that a technological innovation is like a river: its growth and development depending on its tributaries, and on the conditions it encounters on its way. The tributaries of an innovation are inventions, other technologies and scientific discoveries. The conditions it encounters in its development are the vagaries of the market-place.

Semiconductor technology has played its most important role in the development of computers; the history of the development of computer technology since the Second World War is conventionally divided into five generations. The first-generation computers of the late 1940s and early 1950s depended on thousands of thermionic valves or vacuum tubes which were huge, expensive and very liable to malfunction. Second-generation computers were made possible by the invention of the transistor on 23 December 1947, the first small solid-state semiconductor. Solid-state devices are units made of semiconducting material, such as germanium, silicon or the 'crystal' (galena) of early radio sets, linked to electrodes. The input signal disturbs relationships between atoms arranged in a lattice formation inside the semiconductor, thereby performing the same function as the valve. Transistors were able to replace valves at a fraction of the cost, size, energy use and heat generation. The further development of transistors in the 1950s permitted a dramatic reduction in computer size and cost, together with a vast increase in capacity. There were also many improvements in computer storage and retrieval capacity at the same time. By the 1960s transistors themselves had diminished in size a thousand-fold.

Third-generation computers resulted from the development of integrated circuitry in a single manufacturing process. The manufacture of integrated circuits developed between 1956 and 1962 when it was discovered that if a small chip of silicon was suitably treated, it would behave as several transistors, instead of just one. It therefore became possible to build an entire circuit on one silicon 'chip', and these could be made cheaply enough for widespread commercial and military use. Small-scale integration, with up to about 10 transistors on a single chip gave way to medium-scale integration, with up to 100 transistors. By the 1970s, large-scale integration (LSI) technology, with thousands of transistors on a chip, made possible the mass-production of 'memory chips' to store data and 'logic chips' to perform calculations.

The fourth-generation of computer technology can be identified with the development, in the late 1970s, of very large-scale integration (VLSI), in which hundreds of thousands of transistors are embedded on a single silicon chip. The single VLSI microprocessor chip at the core of the Hewlett-Packard 9000 computer contains 450,000 transistors.

An important difference between VLSI technology and its predecessor, is that, at lower densities, circuits could be produced without automated equipment. As recently as 1983, fully 80 per cent of the circuits sold by US manufacturers were assembled and tested in low-technology plants, usually in developing countries, using semi-skilled labour. However, when the circuits are so densely packed that the lines are perhaps two millionths of a metre apart, then VLSI production requires a largely automated process, from the design of the circuits right through to the packaging.

Although integrated circuit technology has not yet reached the limits of its development, there are waiting in the wings even more dramatic computing technologies, based on 'magnetic bubbles' and lasers, which will carry still further the remarkable progress which has hitherto been made in speed of performance, reliability, miniaturisation and cheapness.

It is precisely because microprocessors are so small, so cheap, so powerful and so reliable that they are spreading into every branch of economic activity and social life. The microprocessor is becoming a universal tool, because it makes it possible for unskilled people to perform routinely a large variety of complex functions. A simple example is the pocket calculator, which allows its operator to perform many mathematical functions which he would not otherwise be able to do for himself.

Another application, whose wider importance is just beginning to make itself felt, is to be found in the development of information technology. Information technology is the coming together of three hitherto separate technologies—those of communications, computers and office equipment. At the heart of this convergence is one common factor, the microprocessor. From the user's point of view, information technology means rapid, reliable and low-cost transmission of messages received as sound, image or electronic signals from anywhere in the world to anywhere else. This is combined with a capability for the cheap and rapid storage, reproduction and pre-programmed processing of these messages.

A further development in computer technology is the projected fifth-generation computer. Since it has not yet been invented, we shall not be discussing it in this book, except to comment in Chapter 11 on the research programmes which at this moment are being directed in Japan, the United States and Europe towards its realisation. Even if the objectives of this ambitious research programme should not be realised fully, it seems certain that enough will be achieved to maintain at least to the end of this century the remarkable momentum of progress which has been sustained so far in the development of microprocessor technology.

That technology has already spread rapidly. A 1984 survey of 1200 factories in Britain identified approximately 100 different types of product ranging from fork-lift trucks, pressure gauges, sewing-machines, petrol pumps and cash registers, to washing-machines, toys and medical equipment, which incorporated microprocessors. In the same survey a similar number of production processes were identified as being controlled by microprocessors, ranging from blending, colour sorting and moulding, to laminating, drilling, painting and even pricing. In fact, the existing and potential uses of microprocessors are so pervasive that it is difficult to think of any area of economic activity that will not be affected by it very shortly.

Given that there is such a wide range of microprocessor-controlled and related technologies one must obviously be cautious in generalising about their effects. Some of these technologies have general effects, others are quite specific. Likewise, some human reactions to the deployment of a given piece of new technology are common; others are highly individual. Furthermore, even where a relevant technology is well understood, it is changing all the time and many of its more important consequences may be quite unpredictable. This does not apply only to basic research discoveries, but also to the improvements in new technology which occur once it has been put in place. For instance, automatic cash dispensers were originally intended to be located inside banks: it was only later, when their potential began to be realised, that they were moved first of all to outside walls of the banks, and then to non-banking outlets such as hospitals and supermarkets. This is a contemporary example of something which has long been known to economic historians: namely that increases in productivity can sometimes arise as much from continuous small improvements in the operation of new machinery on the job (learning-by-doing) as from the first application of spectacular new inventions. Thus for example, James Watt's famous invention of the separate condenser doubled the efficiency of then existing steam engines, but in the following 50 years the efficiency of steam engines rose a further fourfold within the basic Watt technology.

New technology is not the only factor bringing about economic and social change at the present time, but it is certainly an important one.

There have already been a number of books on the subject of new technology and its effects on modern society. Some of these, (e.g. *The Computerisation of Society* by Simon Nora) have been extremely distinguished. But most of these works date from the late 1970s, before the effects of the new wave of technology had made themselves felt. Consequently, in a fast-changing situation, many of them are now out of date. Still earlier, but perhaps less notable, studies were remarkable for the dire warnings which they contained of the dehumanising and unemployment-creating effects of new technology. Studies of what has actually happened, including our own, do not bear out these alarmist fears (although it is early days yet). It is true that unemployment has risen dramatically in most of the industrialised countries in the last 10 years, but few of those best placed to judge have attributed this principally to the effects of new technology.

There have indeed been a growing number of empirical studies of the effects of new technology in recent years. They largely take the form of very specific case-studies: they have been well described as 'semi-isolated clusters of facts organised by special purpose theories' (Nelson and Winter, 1982). Of course, all facts have to be organised by some kind of theory, and the facts presented in this book are no exception. It is therefore only right that we should begin by setting out our own view of the nature of economic growth in modern society, i.e. the framework into which the particular ideas expressed in this book are fitted and by which the empirical evidence reviewed and expounded here is interpreted.

Ours is an evolutionary view of the economy and society in

market economy countries, characterised by competitive innovation between firms. There is a continuing qualitative change in products, processes and equipment, and in human tasks and organisations. Progress in these dimensions, like progress in technology itself, is of an uneven, groping character, marked by uncertainty and by the existence of transient gains and losses. The changes which take place are neither continuous nor smooth. Change is disruptive, and a faster rate of change means even greater disturbance.

For example, changes in technology can switch cost advantages between firms and countries in sudden and unpredictable ways. A firm which is put under competitive pressure as a result of a rival's technical advantage, must either reduce its costs (perhaps by changing its internal organisation), or go out of business, thereby changing the organisation of the industry. The disruptive effects of a change in technology are particularly severe for specific forms of capital, and for workers with specific skills. Their earnings can be abruptly reduced, and in the extreme case they can become permanently redundant. Nothing can be done about obsolete specific capital, but labour redundancy is an organisational problem, which we believe is a responsibility for firms, governments and workers themselves to resolve.

Such disruptive effects of change may be less visible, because they are less direct, than the positive gains, but they are no less a consequence. The process of economic growth is therefore a process of 'creative destruction'. For the individual firm in this process there is not one, but many, survival strategies: so the economic world is characterised by a diversity of observed characteristics of firms.

Much has been written on the subject of how new technology moves from being simply an improvement in formal knowledge (invention), to being commerically successful (innovation). This book is not concerned with such questions. It looks at the deployment of new technology from the point of view of the typical firm which does not itself engage in significant research and development expenditures, but which is content to buy new technology 'off the shelf', as it were, when it becomes available on the market. Different firms, of course, react with different speed to the availability of new technology on the market. In our own survey we tried to draw a distinction between those firms which were quick off the mark in adopting new technology (the adaptors), and others which did not avail themselves of the opportunities which their competitors had seized (the laggards).

Whether a firm moved quickly or slowly, all were agreed that the motive which drove them to adopt new technology could be summed up in the single word: competitiveness. The need to keep ahead of, or at least abreast of, whatever one's competitors were doing in terms of the adoption of cost-reducing process technology or quality-improving product technology was cited time and again as the single most important reason for adopting new technology. This confirms the results of other studies which have shown, for example, that new technology is regarded as the 'cutting-edge' of competitiveness in Japan.

New technology can improve the competitive position of a firm in many ways, but most often it does so by reducing the firm's labour costs per unit of output. This can mean a reduction in the number of people directly employed in that firm, whether in factories or in offices or both, although if the firm is able to increase its output sufficiently, as has happened in many fast-growing industries, this need not be the case. To the extent that the firm which successfully deploys the new technology can increase its market share at the expense of a rival which is still using the older, higher-cost technology, then the new technology may indirectly bring about job losses by causing lay-offs amongst the firm's rival and other less successful competitors.

The effects of new technology on overall employment in a country is a controversial and complicated question, which we look at in Chapter 6. Apart from the job losses which new technology brings about, both directly and indirectly, there are of course job gains which can be attributed to it. There are first of all the jobs which are created in order to manufacture the capital equipment which embodies the new technology. Then the higher incomes which the use of new technology makes possible (either in the form of higher wages or higher profits or both) should result in higher consumer and investment spending. But some of these effects are very difficult to pin down, and will vary from one circumstance to another. In one instance, the overall job-creating effects of the deployment of a particular piece of new technology may outweigh the overall job-destroying effects: in another instance they may not. If there are net gains in jobs, many of these may be outside the country. Even if job gains outnumber job losses, there still remains the major problem of moving workers from 'old' to 'new' jobs. This is perhaps the single most important challenge in the field of economic policy facing the governments of industrialised societies at the present time.

While the reduction of unit labour costs is perhaps the most common effect of new technology in production, there are other effects which contribute to the lowering of the firm's total costs or the improvement of product quality or both. These effects include increasing the speed of output, widening the range of output, achieving a reduction of inventory, and therefore of inventory costs, the provision of a faster or more reliable service to customers, reduced wastage, a reduced lead time for putting new designs into production, greater product reliability, and greater flexibility in the organisation of the process of production.

It has often been alleged that the adoption of more and more sophisticated technology in the process of production as well as in the process of consumption will have a stultifying effect on the emotions of human beings. This is most dramatically expressed in such famous works of social science fiction as Aldous Huxley's *Brave New World.* More bluntly, the economist E. J. Mishan wrote in 1969 that 'the technical means designed to pursue further material ends may produce a civilisation uncongenial to the psychic needs of ordinary men'. In our survey, we were therefore sensitive to evidence of such dehumanising effects. It required persistent questioning to discover it, and such evidence as there was was fragmentary and confined to particular individuals. The typical reaction of most workers was one of impatience for the introduction of new technology in their own workplace, believing that it would bring them better working conditions and greater security of employment as well as other benefits.

A related criticism which is commonly levelled at new technology is that it is destroying traditional skills, particularly the skills of the craftsman. We look at the evidence for these charges in Chapter 7. In that chapter we also look at the issue of skill-polarisation, the suggestion that one of the effects of new technology is to create a surplus of unskilled workers and a shortage of highly skilled workers. Whether or not new technology is wholly to blame, there can be little doubt that in the advanced economies of the present day there is a serious problem of skill mismatching. In other words, the skills of those who are being made redundant do not, on the whole, match those required by employers seeking workers for new jobs. New jobs may require technicians rather than craftsmen, or unskilled women workers rather than skilled men.

In this book, we consider the economic impact of new technology on three groups of organisations: firms, governments and trade unions. We consider the effect which new technology has had upon them in Chapters 4 and 5, and we discuss the challenges which it poses for them in Chapters 8–10.

Firms are challenged by the forces of competition to adapt or to perish. Whether they will survive depends in part on their own internal organisation, and in Chapter 8 we deal in particular with questions of internal communications, human relations and training. There are no such urgent pressures on the administrative staffs of government to avail themselves of the cost-reducing and service-improving opportunities created by new technology. Nevertheless, in a democratic society, governments must find a way of bringing about the necessary innovations if consumers of public services are to be satisfied, and if taxpayers are not to be further burdened. In addition to these responsibilities, governments have challenges to face in the field of such policies as technology, employment and training, and temptations to resist, such as increasing their powers at the expense of the individual citizens. These issues are discussed in Chapter 9.

Of all organisations in the modern economy, it is perhaps the trade unions which are faced with the most serious challenge by the advent of new technology: this is discussed in Chapter 10. Since the trade union movement began more than 100 years ago, the craft unions have provided its backbone in most of the industrialised countries. But new technology is eliminating the need for many of these crafts, and the membership of the craft unions is falling, as traditional jobs are being eliminated. Although unions are retaining and perhaps increasing their strength in the public sector of advanced economies, the jobs being created in the private sector, as a result of the spread of new technology, tend to be in occupations which, on the whole, are not strongly unionised. It therefore seems that trade unions, too, must adapt their activities if they are to survive.

These, then, are some of the challenges which arise from new technology. New technology creates new opportunities, but also dangers. What happens to the economy of a country will depend on how firms, governments and trade unions, and individuals within these organisations respond to these challenges. Their response, rather than the technology itself, will be decisive.

2 Competitiveness

If a manager or worker is asked the question: why did your company decide to adopt new technology? nine times out of ten the answer will be, to improve its competitive position. Competitiveness has many different facets, but it can be loosely summarised as any measure which lowers the costs of a company's product or raises that product's quality.

In this chapter, we look at some of the ways in which new technology can influence a company's competitiveness, particularly on the product side. We emphasise the importance of human factors in determining the success or otherwise of the deployment decision, and we suggest that competition by means of innovation is a process which has some obvious benefits to the consumer, but some less obvious disadvantages to some producers. The effects of new process technology are dealt with in more detail in Chapter 3.

New technology improves the competitiveness of firms in different ways according to the specific nature of the technology and the circumstances of the particular firm. Improvements in competitiveness can be divided into four classes: (1) those which lower labour costs, (2) those which lower capital costs, (3) those which improve product quality, and (4) those which extend the range of products or services offered by the firm.

The reduction of labour costs per unit of output is perhaps the most common effect of the introduction of new technology. This does not necessarily mean a reduction in the numbers of people employed in the plant or office concerned. In many industries—banking, insurance and finance are perhaps the most prominent examples—new technology has permitted a faster growth of output, which has meant that lower unit labour costs have been combined with no redundancies. In other industries, direct labour-shedding has so far been sufficiently small that it could be accommodated by 'natural wastage'. In some industries, such as mechanical engineering, the introduction of new technology has apparently been accompanied by massive labour-shedding, especially during the recession years 1980–82, but closer inspection shows that the introduction of new technology often followed a cost-reducing reorganisation of production, rather than accompanied it.

Prior to the advent of new technology, many engineering firms had large amounts of working capital tied up in the form of stocks of finished and semi-finished goods, materials and components. With high interest rates, the costs of holding large inventories can be a significant part of total costs. By dramatically reducing the time required to produce a particular product, and by making possible efficient and automated management control systems, computerisation can significantly reduce inventory costs as well as increasing the firm's flexibility of response to new orders. As an example of increased speed of production we found the case of a particular machine part which had taken roughly 24 hours to produce by manually-operated processes, but which could be produced in 16 seconds on a new CNC lathe. To fulfil quickly an unexpected order, the firm no longer needed to maintain a large stock of parts.

Reductions in equipment costs can be used to offset a competitor's lower labour costs as rival producers battle it out to achieve the lowest possible total costs for the same product. For more than half a century, the textile industries of Western Europe and North America have been waging (and losing) a defensive trade war against imports from the developing countries, whose competitive advantage has been based upon low costs of labour. Now, new technology offers western producers the opportunity of reversing this cost advantage, and the big companies are moving into high-technology spinning, weaving and knitting. If they are prepared to work their new equipment round the clock, seven days a week, they will be able to achieve in many cases even lower costs of production than those in the poorer countries.

A third way in which new technology can improve a firm's competitiveness is through improved product quality. This is not always true: in the printing industry, the replacement of the old technology by the new has sometimes meant a decline in the quality of printing and binding. In most cases, the customer feels that this loss is more than compensated for by the shorter delivery times and by the lower costs which may be passed on to him. In manufacturing generally, however, and especially in engineering, the increased precision and flexibility made possible by microprocessor technology leads to improved product quality. In both engineering and electronics, hitherto impossible shapes, sizes and designs are now being produced.

Finally, new technology has made it possible for firms to improve their competitive position by offering their customers a wider range of products or services than they were able to do before. Computerisation has made possible a major reorganisation in the financial sector, where banks, building societies and insurance companies now vie with each other in offering a complete range of financial services to the customer. Previously each type of institution would offer a single service or at best a limited range of services. At the other end of the scale of size, small jobbing printers who have introduced the new generation of equipment have found that it has allowed them not only to offer a much shorter delivery time to their customers, but has also allowed them to compete for business which was previously beyond the range of their capabilities.

In industries where the technology is changing rapidly and competition is intense, the survival of the firm itself may be at stake. For firms in the electronics industry in the 1980s the adoption of new technology has been essential if a firm is to stay in business. The American semiconductor industry provides an illustration. Rapidly improving production techniques and the entry of new firms have resulted in competitive price-cutting. Companies within the industry have had to learn to live with constant change in order to survive.

As Braun and Macdonald (1978) show, once the transistor was invented in 1947 what mattered thereafter was the ability to make the devices reliably and at reasonable cost. The race was on for the right manufacturing technology, and a highly risky contest it proved to be. Invest too little or in the wrong technique and the firm could fall irretrievably behind. Invest too much and the firm risked committing itself to a particular process well into its obsolescence. The early transistor devices were neither reliable enough nor cheap enough to create markets for themselves easily. They had to seek out small corners of the electronics market in which they could sell (e.g. hearing aids).

However, as the manufacturing technology progressed, the product became stronger and became able to create markets for itself. The types of device which were made were dictated by the ease of manufacture, i.e. low cost. Because active semiconductor devices are very cheap in terms of cost per active circuit component, the point was reached that not only could integrated circuits create markets for themselves, they also began to alter technologies in related fields.

Ironically, the very industry which has made possible so much automation in so many other industries has itself experienced relatively little. In the microelectronics industry, products and processes have changed so rapidly that automated production lines have generally become obsolete before they have recovered their cost. Most companies, especially the larger ones, have wanted to automate their production lines, and many have tried at different periods in the past when they felt that the technology was becoming stable. Such attempts have nearly always proved to be expensive failures.

Peter Drucker (1985) has studied the development of the computer industry, and he believes that the experience supports the proposition that the higher the technology the higher the risk. He suggests that innovations based on new technology follow an identifiable pattern. For a long time there is an awareness that an innovation is about to happen—but it does not happen. Then suddenly there is a near-explosion, followed by a few short years of great excitement, lots of start-up activity and tremendous publicity. Five years later comes a shake-out, which few firms survive. In high-tech innovations, time is not on the side of the innovator.

There is only a short period—a window in time—during which entry is possible at all. Usually, innovators do not get a second chance: they have to be right the first time.

According to Drucker, the first window in the computer industry lasted from 1949 to 1955. During this period every major electrical apparatus company in the world went into the manufacture of computers. By 1970, every one of these major companies had ignominiously retired from the field, which was left to companies that had either not existed at all in 1949 or had then been small and marginal. In the late 1970s, a second window opened with the invention of the microchip, which led to the development of word processors, mini-computers, personal computers, and the merging of computer and telephone switchboards. The companies that failed in the first round did not come back in the second. But because of the emergence of a world market and of global communications, the number of new entrants to the industry during each window has greatly increased. When the shake-out comes, if the number of surviving firms in the mature and stable industry is as large as before, the casualty rate is going to be much higher, says Drucker. Already a shake-out has begun among microprocessor, mini-computer and personal computer companies. Today there are perhaps 100 such companies in the US alone: by 1995 Drucker predicts that there are unlikely to be more than a dozen of any size or significance left in the world.

Which firms survive in the computer industry or in any other industry is impossible to predict. There is no unique prescription that will guarantee survival, and there would appear to be many diverse strategies which are open to the successful firm. Some of these strategies are competitive, aiming at market or industry leadership, if not at dominance. Other strategies are aimed at making its successful practitioners immune to competition and thus unlikely to be challenged. As in biological evolution, chance plays an important part. Competitive strategies which have been successful in one country may not be successful in another. The growth of the semiconductor industry in Japan has been quite different from what has happened in America. In Japan, growth came primarily from the established valve producers, with extensive government interference, with virtually no labour mobility, and with no military market.

It should be emphasised that the adoption of new technology is not the only factor in a firm's competitive armoury. On the contrary, it may be just one element of an overall cost-reduction and reorganisation package. Nor should it be supposed that because new technology represents a step forward in scientific or technical capability, that the decision to deploy it is thereby made on 'rational' or 'scientific' or 'objective' grounds. We have already seen that the more advanced the technology the riskier the investment decision.

Furthermore, the decision to deploy new technology depends not only on changes in objective circumstances, but in the perceptions of businessmen. After all, there is often a long lead time, sometimes as long as 25 years, between the first availability of a technology and its commerical deployment. IBM's personal computer has not captured a large share of its market for any 'objective' reasons: quite the reverse. It is more costly than, and technically inferior to, many of its rivals. It is bought because of a subjective judgement about the reliability not only of the product but even of the company itself. No purchasing manager ever got fired for buying IBM equipment. Yet a discounted cash-flow calculation, or similar 'objective' investment criterion would not lead to the IBM machine being chosen.

Another subjective factor which enters the competitive process is prestige; the prestige that comes from being the first firm in an industry to deploy the latest technology, or at least from enjoying the reputation of being a firm which is always in the forefront of technological advance. Although there are many firms who say wisely that they would never deploy the latest available technology just for its own sake, there can be no doubt that among larger firms the reputation of always being ahead of one's rivals in the deployment of new technology exercises a powerful stimulus to management and employee morale. Of course, this sort of competition is not confined to business. As Braun and Macdonald point out, one of the main stakes in the scientific game is prestige: the prestige that comes from being the first to articulate a new idea, to find a new formula, or get a new experiment working.

Whatever the particular features which characterise competition through innovation in different industries and in different countries, the consequences are always the same. Within an industry, there are falling product prices and increasing output volumes. At the level of the firm, those firms which innovate successfully survive and flourish, for a time at least. This success is, however, obtained at the expense of their less fortunate rivals. Those firms which do not innovate, or which innovate unsuccessfully, decline or fade away completely.

Those industries where the new technology is embodied in the products provide some of the most remarkable illustrations of the price-reducing effects of innovative competition. A 1 kilobit chip cost $1280 in 1975; by 1985 the price had fallen to less than $100.

Such a reduction in price, to say nothing of the simultaneous improvements in quality and performance, has meant that the cost of products incorporating such electronic components has also fallen dramatically. The computing power that cost around $15 million 30 years ago can today be obtained for under $1000, after adjusting prices for inflation. And the rate of decrease in prices continues unabated: today's personal desktop computer is about ten times cheaper, in real terms, than its 1975 equivalent.

Another example of falling price is provided by the hand-held electronic calculator. Mechanical calculating machines were available in the 1930s at about $1200—then the price of two motor cars. By the 1950s, electro-mechanical calculating machines were being made for perhaps half that sum. The first electronic calculator was produced in 1963, and by the second half of the 1960s calculators using integrated circuits were available. Within a five-year period in the 1970s, the price of the simplest electronic calculator dropped from around $100 to about $15. Today it is less than $10.

A still more recent product of the semiconductor industry is the electronic watch. In 1973, there were about 250,000 electronic watches produced, virtually all in the US, selling for an average price of $250. By 1975, production had reached 3 million at an average price of about $150. By 1985, world production of electronic watches exceeded 50 million, at an average price of less than $30.

It is easy to identify these and other benefits which flow from the introduction of new technology. The new products and the new processes spell new jobs and new profits in the successful firms, and a net gain for consumers overall. These are all the creative effects of technical innovation and they are immediately recognisable. It is not so easy to identify the destructive effects of technical innovation.

It is perhaps not difficult to imagine that new products take markets from existing products, making redundant the services of the workers who are engaged in that production and rendering obsolete the capital which is similarly engaged. In the same way, new processes of production undercut existing technologies used by rival firms, causing them to make losses, to lay off workers and close down capital, and ultimately driving the marginal ones out of business. But because all of these effects are indirect, it is less easy to see the connection between the deployment of new technology in a particular plant, and the destruction of capital and of job opportunities elsewhere. In the following three chapters we shall look at these effects in further detail. In the meantime, it is sufficient to state that the introduction of new technology sets in motion a process of creative destruction, to use Schumpeter's phrase. There are losers as well as gainers in this process, and changes in technology, wherever they occur, can cause severe dislocations in industry, the closure of factories and redundancy or redeployment of labour all over the world.

The more specific the capital, the more radical the nature of the new technology, and the more rapid the rate of change, then the greater will be the consequent dislocation. It will affect not only the producers of existing products and the users of existing processes. They will lose markets and find their costs of production undercut; it will also affect the unsuccessful competitors in the process of innovation. It has been estimated that it took 30 years from the start of the computer industry in the late 1940s for that industry to reach break-even point. Until the early 1980s the profits of the few successful computer manufacturers were more than offset by the enormous losses suffered by the rest, principally the large international electrical companies, such as GE, Westinghouse, RCA, GEC, Plessey, Ferranti, Siemens, AEG and Philips, all of whom ultimately failed in their attempt to remain major computer manufacturers.

What happened in the growth of the computer industry differed only in detail from what happened earlier in the introduction of new technology in other industries in the western world. A similar pattern of competitive innovation can be discerned in the growth of railways in the early nineteenth century, and in the expansion of the electrical appliance industry in the 1920s. In each of these earlier waves of innovation there is a clear pattern of cause and effect from which we can learn. The present wave of innovation is however taking place in a very different social and political environment from that which has gone before, and therefore we should expect to find differences as well as similarities in the effects of the introduction of new technology today.

OTHER INFLUENCES

Although competitiveness is the most widely quoted reason for the deployment of new technology, it is not the only one. Survival was also mentioned in our survey on a number of occasions. Although it can be regarded as simply an acute stage of competitiveness, in some cases company survival is a more fundamental matter than product competitiveness. For example, survival for firms in the electronics industry usually means the ability to break into a new product range. Where existing products are becoming obsolete it is essential to change with the market or go out of business.

In many cases companies invest in leading edge equipment in order to be capable of reaping the potential benefits these machines can bring—even if they are unforeseen at the time of purchase.

In banking, insurance and finance one reason for deploying new technology was of even greater importance than competitiveness. This was the ability to cope with the increasing volume of work: computerisation also enhanced the firm's ability to improve customer services.

Other reasons were given for the decision to deploy new technology. These include the availability of investment grants provided by the government to encourage regional development, and the need to replace old equipment. A further reason given was that new technology provided the opportunity for greater management control.

A business recession—whether general, or confined to the particular markets a firm is serving—can play a key role in influencing the timing of a decision to deploy new technology. Very often, it provides the final incentive which pushes the firm into making the decision about which it had been hesitating. A recession sometimes provides management with a further incentive to invest, in the shape of a shift of bargaining power in favour of management. This is based upon the realisation by the workforce that few alternative job opportunities are available.

NON-DEPLOYMENT

Of course, not all firms decide to deploy new technology when it becomes available to them. The main reason is, quite simply, that of cost. There is another factor common to most non-deploying firms: these firms tend to be small, with a narrow niche in their own particular market, which, in some cases, just enables them to survive and no more. Where this represents highly specialised production the pressures to deploy new technology tend to be less acute.

Other reasons commonly given for non-deployment included:

(a) The product consisted of a large variety of small batches (new technology still being considered more efficient on large or standardised production-runs).
(b) The suggestion that quality variations in material were not easily handled by automatic machines.
(c) The rate of prospective change of technology was so rapid that any investment now would be undermined by a still lower cost process within two years.
(d) Seasonal work made intensive use of the machinery impossible (although it was often the case that new technology could have widened the product base if applied with imagination).

However, these may be excuses rather than reasons. As a recent British government report (ACARD, 1983) comments: 'Too many companies have not yet applied advanced technology to their manufacturing process.... Those companies will become progressively less competitive and many will not survive the next ten years.... In our view, the principal cause of failure to invest is a lack of conviction of the vital important of doing so.'

It must be emphasised that any piece of new equipment embodying a certain level of technology does not automatically confer a certain cost or performance advantage upon the firm which deploys it. That depends on the way it is used. Japan's success in attacking the US car market has not been brought about by any significant technical advantage either of the product or the way it is produced. Although they have earned a reputation for reliability, most Japanese cars are fairly ordinary in a technical sense. Nor has their apparent cost advantage been due hitherto to any superior degree of automation of their production lines, compared to their US competitors. As Hayes and Abernathy (1980) point out, the Japanese productivity record and cost advantages simply boil down to the human relationships between the employee and the organisation for which he works. It is not a coincidence that in their well-known book about America's best-run companies, Peters and Waterman (1982) identify eight attributes of management excellence, and every one of the eight is about people.

The lesson to be learned from Japanese and US industrial experience is that the application of new technology, like any other facet of competitiveness, will be successful or unsuccessful depending on the human factor: on communications, training and basic human relations (i.e. respect for the individual), within the organisation making the investment. Consequently, it is these factors which we dwell upon in the following chapters of this book.

PART II
Effects of New Technology

3 Material Effects

This chapter deals with all those effects of new technology other than its effects on human beings and their organisation. In other words, it deals with the effects of new technology on capital and on output. We begin by describing the kinds of new technology which are applied in a typical manufacturing industry—engineering. Then we identify some of the effects which new technology has on the performance of activities within the factory or the office and on products and services. Finally, the indirect effects on capital and output in the rest of the economy, arising from the deployment of new technology in some industries, are discussed.

Engineering can be divided into three broad functions: design, production and administration. Design involves the preparation of engineering drawings by draughtsmen. Production consists of a sequence of processes such as machining, handling and assembly. Administration consists mainly of maintaining records of orders, arrivals of materials and despatches of finished goods, levels of stocks and work in progress, and payment to employees. All three of these basic functions are gradually being computerised, and one day they will be joined together in a single computer-controlled operation in a totally automatic factory.

The totally automated factory is still in the future, although not far away. For the time being, the advances in manufacturing technology which are actually at work are dramatic enough. Amongst many other advantages which they confer on the firms which adopt them are the ability to produce a wide variety of batch items at costs approaching those of volume production, with reduced manufacturing lead times, improved quality and reduced inventories. The following are some of the more important technologies now in operation in the engineering industry in the industrialised countries.

DESIGN

Engineers and draughtsmen use computers to design products, and to work out how they will be manufactured. There is a family of computer-based facilities, using specialised software and data files, for the design of components and assemblies, the preparation of diagrams and perspective views, engineering calculations, and standardisation of components and tooling. These facilities have the collective name of computer-aided design and draughting, or CAD for short.

PRODUCTION

Design and manufacture are linked by computerised systems which define operating sequences, create control tapes for CNC machines, and establish requirements for fixtures and tooling. These systems are known as computer-aided manufacture, or CAM. Computer numerically-controlled machine tools (CNC) are individual machines which use computer controllers to store and perform operating instructions, such as the selection of cutting tools and speeds, with manual loading and supervision. Such machines can process a sequence of different batches with low changeover times between products.

In flexible machining or manufacturing systems (FMS), two or more machine tools are controlled by a supervisory computer which also arranges for the blocks of metal being machined to be moved from one tool to another. The method of transport may be by conveyor belt, by a series of robots or by unmanned trolleys. The computer not only instructs the machine tools to execute a wide range of functions, it also directs the transport mechanism to operate in the most efficient manner, to achieve for example maximum spindle utilisation or minimum lead time.

The functions of production and administration can be linked by a computerised manufacturing control system which combines a database-handling operating sequence, shop loading and scheduling, and despatch procedures as well as monitoring work-in-progress, with a computerised inventory control. More advanced systems can include product-costing.

ADMINISTRATION

A computerised inventory control system can handle bills of materials, inventory recording and control and purchase ordering and control. There are also many other systems for dealing with sales, deliveries and invoices, for accounts and for the payment of wages. There are even computerised systems for stores, which can control the movement of parts and materials to storage locations, and their later delivery to workstations.

Each of these technologies brings its own benefits, and these benefits can increase with integration (for example CAD with CAM). But the effects will depend on the sequence of integration and the operations of the individual company. The sequence in which each company will introduce new technology will depend on its individual situation and the current level of its technology. However, a typical engineering company would probably begin by introducing inventory control, the first half of a totally computerised production control system. A one-off reduction in inventory would give an immediate cash benefit, with further benefits being gained as inventories gradually slim down further over two or three years.

The changeover from the old manual to the new computerised system can be eased by introducing the new system in stages, one module at a time, with different modules covering different tasks, such as bills of materials, purchase ordering and other activities. The inventory control part of the system might be brought into use over a two-year period, followed by the second half of the system, i.e. computerised production control with a modular approach again being adopted. In due course, however, all the modules should come together to form an integrated inventory and production-control package. This package should be at the heart of any manufacturing company's programme of new technology, giving flexibility of control and speed of response. A recent survey by the British government's National Economic Development Office (1985) has shown that the deployment of such new technology can reduce a typical engineering company's material costs by around 14 per cent.

In general, the principal direct material benefits to a firm introducing new technology include one or a combination of the following:

(a) a reduction in the levels of stocks and work-in-progress,
(b) faster rates of output,
(c) improvements in the quality of the product, and
(d) greater flexibility in production.

The greater part of a reduction in inventory costs comes not simply from improved methods of inventory control but from the greatly increased speeds of output which are made possible by new technology. For example, a CNC machine can reduce the time taken to machine a complex part by as much as a factor of 1000. This means of course, that to satisfy any given order, a greater proportion of any delivery can be 'made to order', diminishing the need for a large inventory of components, raw materials or finished products, all of which are extremely expensive in terms of control costs. Such dramatic improvements in the rate of output are not confined to engineering, however. The new technology can deliver equally high speeds in printing. For example, in the printing industry, typesetting has progressed from manual keying, which releases the metal type into the correct position on the block, to laser composition. The number of characters per hour which can be set in type has increased from a few thousand to 1½ million with the latest laser equipment. Less spectacularly, an insurance company can, with computerisation, reduce the time it takes to issue a policy from one week to two days. Likewise, a computer can update a customer's account in a bank or a building society in a matter of seconds rather than minutes.

In these latter two cases, the advantage of greater speed of output is not reflected in lower inventory costs, but in greater customer satisfaction, as well as in lower labour costs, thus giving the firm which deploys the new technology a competitive edge over those which have not. In printing, in the financial sector and in other service sectors, speed of service to the customer is critical to competitiveness, and a faster service can be likened to an improvement in product quality in the manufacturing sector. While there is no doubt that the manifestation of new technology in the form of automatic cash dispensers, front office terminals and computerised customer records has improved customer services in the financial sector, the quality of service in this sector also includes such aspects as the level of courtesy displayed or the ability to devote time to the problems of each individual customer. It was apparent from our own survey that, while the enormous improvement in speed of data processing in the financial sector has released staff time to deal with the individual problems of a greatly increased number of customers, there is some evidence of an overall decline in standards of courtesy in recent years. However, it may be unfair to attribute this entirely to new technology.

There are other cases in which the adoption of new technology has been accompanied by a noticeable decline in quality. Examples can be quoted from three industries: printing, food-processing and clothing, although in only one case was it clear that new technology was the cause of the decline in product quality. Craftsmen in the binding departments of some printing companies lamented the fact that lower standards of work formed an increasing proportion of their work as compared to three decades ago. However, they recognised that customers were, in general, no longer prepared to pay for quality binding, and preferred the lower quality product which was associated with a reduced cost. In this case, therefore, it would seem that the reduction in quality is brought about by changes in consumer tastes rather than by technology.

A chocolate manufacturer had abandoned hand-decorated confectionery and substituted vegetable oil for the more expensive cocoa butter at a time when computer-controlled equipment was being introduced into the factory. On closer examination it appeared, however, that these quality-reducing changes were not a necessary consequence of the introduction of the new technology, but were merely a part of a general reorganisation programme within the plant to increase volume and cut production costs—a programme in which the deployment of the new technology played only one part.

In the case of a shirt manufacturer, however, it was clear that new microprocessor-controlled sewing machinery was unable to provide the same quality of finish as had been achieved under the manual method of production. In both these two cases we were assured that while the decline in the quality of the product was evident to a specialist, the great majority of customers would be unable to detect any difference in the product.

In the engineering industry the dramatic improvements in precision achievable by CNC machinery can lead to companies which deploy the new technology being able to extend their range of products to take on work which was previously beyond their capability. One mechanical engineering firm we interviewed was able to manufacture turbine blades of a completely new type, because its new CNC machinery allowed much more complex and accurate machining to take place. Likewise, several printing firms were able to move into entirely new areas of business simply because the new technology allowed them to consider completely new product lines. Banks and other financial institutions have expanded the range of services each offers to their customers. New technology has enabled such changes to occur, rather than instigating them. The ability of other financial institutions to compete with banks in the personal loan market has been greatly enhanced by new technology. This is not only because of the time saved in performing routine tasks, but also the savings in staff training time. Computerisation has made it easier to transfer staff between departments in response to shifting demand, because most of the technical details are handled by the computer. The most striking example, however, of the capability of new technology to widen a firm's possible product range comes from the electronics sector. The design of the complex printed circuit boards (PCBs) used in the industry would be virtually impossible without the use of sophisticated CAD systems, and the assembly of computer components in sufficient quantity to be viable in the market depends in part on the use of items like automatic PCB populators.

Automated manufacturing and the use of computers in design and production management is not new. But the older technology had always lacked flexibility, at least at a reasonable cost, and so its application was restricted to volume production of a single product or to a very narrow range of products. With the new technology, flexible automation is at last commercially viable: it is possible to achieve high-volume productivity with medium-volume, wider-variety output. It is precisely the medium-volume, medium-variety batch producer which counts for about 70 per cent of engineering output in a typical industrialised economy. Until now, such engineering companies have responded to competitive pressures by trying to move towards higher volumes and reduced product variety, in order to be able to use mass-production machinery.

The engineering market today is an international one, as buyers are increasingly willing to re-evaluate their existing suppliers. In this situation, engineering companies are finding that they must respond more quickly to changes in the market. Not only are buyers more demanding in the type of product they require, but manufacturers no longer have as much time as previously to decide whether they wish to compete in a particular product area. In addition, large customers now expect their component suppliers to respond as flexibly and quickly, and with the same quality of effort, as they themselves have to do.

The new technology which is available enables companies to address two problems: to decide quickly whether they want to compete and, if so, how to compete efficiently and flexibly. Apart from the basic manufacturing technology which enables a company to produce small batches and to customise products sufficiently, the benefits of new technology extend to the whole tendering process. Computerised systems give accurate records of manufacturing costs, and computer-aided design systems can produce tendering drawings, both contributing to a more effective overall response to the customer's requirement, by reducing tendering and manufacturing costs and lead times.

The flexible nature of the new technology means that in many cases it can be less specialised than the capital equipment which it is replacing. Although computer-controlled machine tools and industrial robots can be designed to perform one single task or a limited range of tasks, they can also be designed to perform, equally efficiently, a variety of tasks. This phenomenon has a wider significance, since it is evidence of a break in the apparently long-standing tendency towards a greater and greater specialisation of capital equipment as the division of labour increases. It does remain true, however, that very small companies may find it costly to deploy new technology. For them, it can be inflexible, not in the tasks that it can perform, but in the requirement that there must be a high rate of utilisation if it is to pay for itself.

INDIRECT EFFECTS

In deploying new technology, as in the other strategies which it adopts, the typical firm is seeking to lower its overall unit costs and/or enhance the competitiveness of its product, thereby increasing its profitability. But the new technology must be embodied in new capital equipment, so that an increase in capital costs may be incurred, at least temporarily. The firm must hope that this will be offset by reductions in costs such as those we have been discussing, together with reductions in labour costs. From the firm's point of view, these are the only changes that matter: if the effect of introducing the new technology is to leave the firm's competitive position more secure or more profitable than would have been the case without its deployment, then the firm will be well satisfied with its decision.

But the changes brought about within the firm itself will have repercussions outside the firm, which we must try to account for if we wish to look at the effects of new technology on the economy as a whole. Such indirect effects can be divided into two classes: negative and positive.

Negative indirect effects include the consequences which a new production process in one firm may have for its competitors using existing, older, technology. The new process is likely to undercut existing costs of production, thus rendering obsolete the capital equipment in which the older technology is embodied, causing rival firms to make losses, and perhaps driving the more marginal of them out of business altogether. If the new technology takes the form of a new product, or a qualitatively improved existing product, then this, too, is likely to take market share from existing products, as well as from other, less successful, would-be innovators of rival products. The demand for these products will fall, thus rendering the services of capital used in their production redundant.

Positive indirect effects arise because new technology requires new capital equipment, and thus stimulates activity in the capital goods-producing industries. Additional demands for capital are created to the extent that the new technology is more profitable than the old, and that there are further multiplier effects arising from the increases in real income to which this increased profitability gives rise. However, it should be remembered that the market for capital equipment embodying new technology, especially machine tools, is a world-wide one. Whether an additional demand for new capital equipment can be satisfied by suppliers in the same country will depend on how competitive that country's own producers of capital equipment are.

Much attention has been given to the question of whether the negative effects of new technology (both direct and indirect) outweigh the positive effects (both direct and indirect) in terms of incomes. Is the net effect of new technology to increase or to lower the real income of the community in which it takes place? Like so many questions in economics, this is not one to which a conclusive answer can be given, since the effects are impossible to identify in practice. We can only arrive at an answer by theoretical reasoning. We can be fairly confident that, in general, the deployment of new technology increases profitability for the firms which undertake it, otherwise it would not be undertaken. We can also be fairly sure that the deployment of new technology, in process form, reduces labour cost per unit of output, and therefore total cost, since labour costs are after all one of the more important elements of total cost. It is less certain however that new technology will invariably lower the total wages bill of any firm in which it takes place. Even if it does, it is not clear whether the increase in profitability will be greater or less than the fall in the wages bill. If the fall in the wages bill is greater than the gain in profitability, there remain the indirect effects, both positive and negative, to be taken into account. The problem is that these indirect effects cannot be satisfactorily measured, since at more than one stage removed from the initial change (the introduction of the new technology) which sets them in motion, these effects are indistinguishable from the indirect effects of the many other changes which are going on at the same time.

There are, however, two reasons for thinking that in most circumstances in a typical contemporary advanced economy the net effect of the deployment of new technology will be to raise real incomes. The first of these arguments is a theoretical one. The second is an historical one.

If the net effect of new technology within the firm is to increase value added (i.e. the increase in profitability is greater than the fall in the total wages bill), then Say's law suggests that the ultimate effect will be to increase the real income of the economy as a whole.[1] Say's law states that an increase in real income will eventually filter through an economic system until there is an equivalent increase in expenditure: more simply stated, it implies that supply creates its own demand. In the case of a firm deploying new technology, the higher value added which accrues in the first instance as increased profits to the firm must eventually be spent either on the purchase of capital equipment or on higher wages or other factor payments, or in taxation.

The second reason for believing that the net effect of new technology in contemporary advanced economies is to increase the real income of the whole community is that those periods in history which have been notable for technical progress have also been marked by increases in real income, and indeed in employment. A population ten times that of 1700 now lives in Great Britain, at living standards which are immeasurably higher than they were before the Industrial Revolution. As Musson (1982) points out, the increase in population may be regarded as a response to industrial growth and increasing employment opportunities. Between 1851 and 1951, the total working population rose from 9.3 million to 22.6 million. Even in the industries that were mechanising most rapidly, such as textiles, the total numbers employed rose from 1.3 million in 1851 to 1.5 million in 1911, while in mining, where there was also mechanisation, employment rose from under 400,000 in 1851 to over 1.2 million by 1911.

In all those countries in which it occurred, the industrial revolution of the nineteenth and early twentieth centuries has not failed to increase real incomes and employment. More recently, the decade between 1974 and 1984 saw an unprecedented rate of new employment creation in the US economy. This was also a period when the US economy was experiencing a rate of technical progress which was exceptional, both by its own historical standards, and by those of other contemporary advanced countries.

There is a significant difference between the effects of new technology on labour and on capital. If the introduction of a new production process, or a new product, in one firm leads to a worker in another firm being made redundant, then that worker can in principle move to another job. But in a modern advanced economy, industrial capital is characterised by a high degree of specificity. Once a piece of capital has been formed, it cannot in general be adapted to any other task. If it becomes cheaper or more convenient for people to travel by bus than by train, railway carriages and locomotives cannot be turned into buses. If a rise in the price of aviation fuel renders aircraft with new fuel-efficient engines significantly cheaper to operate than aircraft with conventional engines, then the latter will be rendered quite unusable, unless there is a fall in fuel prices. Specific capital cannot migrate from one function to another: it is therefore destroyed by such changes as the advent of new technology. It is of course not physically destroyed, it lies about in an unused state. We have seen in the last decade idle refineries, manufacturing plants closed which will never open again, fleets of idle tankers and squadrons of redundant airliners. In an industrialised economy, when an increasing part of its industrial capital lies unused, this may be an indication, not of the stagnation of that economy, but on the contrary it is evidence of its rapid rate of progress.

Although labour can move from one occupation to another, upon being made redundant, there are barriers to mobility which are often greater in an industrialised economy than in a developing one. A substantial proportion of the workforce being unemployed in an industrialised country may be a sign of stagnation, if there is a general insufficiency of aggregate monetary demand. On the other hand, if there is not, then such unemployment may be, like the unemployment of capital, a sign of rapid economic change.

In this chapter, we have looked at the effects of new technology on the production of material goods and services and on the demand for capital equipment. In the next chapter we look at the effects which the introduction of new technology has on the organisation of firms and industries. It should never be forgotten that the performance of firms adopting new technology depends on how well they implement it, and not on the technology itself.

NOTES

1. While Keynes was probably right in thinking that Say's law does not hold in the short run, it is unlikely that even he would have argued that it does not hold in the long run. After all, he was careful to place his theory in a purely short-run context.

4 Effects of New Technology on Firms and Industries

This chapter describes the effects which new technology has had on the organisation of firms and industries. It is divided into two main parts. In the first part the effects of new technology upon the organisation of work *within* firms is discussed. In the second part, the effects on the organisation of work *between* firms (i.e. within industries) is discussed.

THE ORGANISATION OF WORK WITHIN FIRMS

The first section on the effects within firms is further subdivided into three categories: (a) the effect of new technology on the organisation of individual tasks; (b) the effect on working practices; and (c) the effect on administrative centralisation and decentralisation.

(a) The Organisation of Individual Tasks

New technology has an enabling effect on the organisation of work within firms, not a determining one. It widens the range of choices open to management about the organisation of work. In particular, it allows management to choose whether and how responsibilities should be redistributed. In any redistribution there are bound to be some losers, and in manufacturing it seems that the losers are often numbered amongst those who occupy supervisory or junior management functions in the traditional hierarchy. Nevertheless, surveys of what has actually happened show that the desire to reduce costs does not allow us to predict any particular outcome.

Experience differs from one industry to another, and, within any given industry, from one firm to another.

Our own survey showed that the factors affecting the organisation of individual tasks within a particular firm could be allocated to one of three categories: (a) technology, (b) organisation, and (c) market imperatives. The influence of technology on job flexibility is most clearly seen in the contrast between the experiences of introducing new technology into the printing and the mechanical engineering industries.

In the past 20 years, few industries have experienced such a rapid change in technology as has the printing industry. Since 1970, the methods of printing which have been in use for centuries have become obsolete, and computer technology has taken over. Not only are the initial stages of typesetting and printing now almost completely computerised, but finishing operations also use microprocessor technology to control folding, stitching and binding. While this has greatly altered individual tasks within the industry, it has done nothing to break down the barriers between the principal tasks (compositing, printing and binding), which traditionally make up the activity of printing, although there is some evidence of increased job flexibility within each of these tasks. The situation in mechanical engineering is quite different. There, technology has frequently been introduced into a machine shop with the specific object of increasing labour flexibility, since this is essential for the new technology to be cost-effective. The use of flexible machining centres is designed to maximise the flexibility of both the machinery and the operator; one is of little use without the other.

Why should the effect of the new technology on flexibility have been so different in the two sectors? In mechanical engineering, jobs such as milling, boring and turning, while by no means identical, are sufficiently similar to be within the scope of a given trained worker. Functional specialisation in this industry has largely been determined by the traditional technology, which tied one man to one machine and encouraged a proliferation of trades. The new technology has been able to remove the technological block to mobility between similar tasks within mechanical engineering. In printing, however, the jobs of different departments are too dissimilar to allow much flexibility even where new technology is in use within individual departments.

In the finance sector it was observed that while there had been dramatic increases in the volume of output and in the range of customer services, in two cases the deployment of new technology had not brought about the expected improvement in job flexibility. On closer examination, it turned out that the organisational structure of the company was responsible for the lack of flexibility. In one case, a finance company anticipated that increased staff flexibility would be among the benefits achieved from a move away from paper-based customer accounts at its head office towards records stored on a central computer and accessed by a network of VDUs. While the desired benefits in terms of improvements in speed and customer service were indeed achieved, the improvement in flexibility failed to materialise. The reason lay in the way in which the structure of work was organised; each clerk was allocated to a specific region of the country, which included a given number of branches. The clerk's relationship with these branches was an important element in determining the smooth running of the total operation. This remained just as true of the new system as it did of the old, and the use of the new technology could therefore do nothing to break down what was essentially organisational rigidity.

In the second case the use of new technology actually reduced the level of flexibility within the department. In this company, an army of clerks spent much of their time updating customer record cards by hand, but were expected to perform a range of additional general clerical duties. When a computerised system was introduced into the branch network each clerk was initially allocated a specific VDU with which to perform the updates. This immediately caused a problem because the clerks became possessive about 'their' VDU, and became less willing to leave their desk and perform the more general duties around the department. This problem was resolved in later installations by placing all VDUs in a bank against one wall, any one of which could be used by a clerk at any particular point during the day. Only by doing this could the previous level of flexibility within this department be re-attained.

These examples illustrate the point that in order to realise the full benefits of new technology it may be necessary to alter an existing organisation. A further illustration of the same principle comes from one of the major oil companies which is in the process of computerising the distribution of its product at the retail level within a European country. In its first attempt at computerisation seven years ago, the necessary programs were written specifically to fit the company's existing operations and organisational structure. In 1986 it embarked on a second, more advanced, stage of computerisation, in which it was prepared to adjust its actual existing system of ordering, invoicing and supplying to conform to an off-the-shelf computer software package which it bought from a California software house. The logic behind this decision is that the organisational structure implicit in the package represents better (i.e. more profitable) practice than does its existing method of operation. Few companies are likely to be as far-seeing as this.

One of the sectors in which labour flexibility is most marked is electronics. Many of the firms we interviewed in this industry depended on an assembly-line type of operation in which few of the jobs were intrinsically skilled. The element of flexibility was achieved by having each worker ready and able to perform any one of a wide range of small repetitive tasks. In its more advanced form, all shopfloor employees bore the title 'process operator' and could be moved to any job within the plant at a moment's notice. At first sight it might seem that the reason for this high degree of flexibility lies in some attributes of the companies themselves. Many of them are small, often dependent for their existence and success on one or two individuals, filling some small niche in the market, and above all able to become adaptable in a rapidly changing market. In fact it is this final point that is the key: because the organisation has to be flexible to survive, so too does the workforce. In electronics, the impetus towards flexibility comes neither from the nature of the organisation nor from the technology, but from the imperative of having to survive in a rapidly changing market.

(b) Changes in Working Practices

It is widely believed that a major investment in new technology must result in a movement towards increased shift working, partly because of the need to recover the costs of the capital investment, and partly because of the perceived trend towards machine-paced work which new technology is sometimes thought to encourage. This belief was widespread amongst the workers that we interviewed; and even where no such changes had taken place and where management insisted that there were no such plans, nevertheless a change towards more shift work was often felt by the workforce to be 'just around the corner'. Although in our survey we did find some examples of changes in shift working following the deployment of new technology, no systematic pattern was discernible. In one mechanical engineering company the introduction of CNC machines coincided with the move from two to three shifts, although the general view in the company was that CNC had reinforced the move rather than caused it. Another firm in the same industry had moved form a straight day-shift system to two shifts. While one company in the textiles industry had introduced a two-shift system as a direct result of the deployment of new technology, another company in the same industry had, through the use of multi-manning, reduced the number of shifts from three to one. Some changes in shift patterns also took place in the electronics and printing industries, again with little apparent pattern. The shift issue did not arise in the finance sector, but there were instances of changes in the pattern of the working day as a result of new technology. For example, after installing computerised counter terminals which allowed cashing-up and balancing at the end of the day to be done much more quickly, the branches of one building society began to stay open 30 minutes later.

One tendency which was evident in all five industries that we surveyed concerned the length of the working day. The effect of new technology in reducing overtime was widespread. Employee reaction to the reduction or abolition of overtime was mixed, depending on the circumstances of the particular individual and the way in which overtime had been organised. Where overtime was optional the extra earnings were generally welcomed, and reservations were expressed about their reduction or termination. However, in other instances there was a distinct feeling of relief at the ending of what was effectively compulsory overtime.

There are of course other changes in working practices apart from shift working and overtime. An interesting example comes from the head office of a finance company which was accustomed to sending out several hundred letters a day to its customers. Under the old technology, clerks dictated letters onto cassettes, which were then typed by a dedicated typist before being sent back to the clerk for signing. Because the standard of dictation and of typing varied so greatly, management wanted a system which would allow better monitoring and control of these functions. The new technology allowed them to introduce a centralised dictation system according to which letters are dictated via the internal telephone to a general typing pool. Dictation and typing are now automatically logged on a computer. As a result, productivity, measured in lines of typing per typist per day, rose by 150 per cent, but at the cost of a significant decline in relations between management and staff, and in the morale of both clerks and typists. The new organisation of work is disliked by the clerks, whose dictation must now be more precise, and by the typists, who dislike the idea of their relative productivity becoming common knowledge.

One of the curious features of our survey was the general tendency for interviewees in all industries, and in all types of work, to play down the significance of changes in working practices. Indeed, there was even a tendency for people to claim that no changes in working practices had occurred, and only to admit under close questioning that some had taken place. Three reasons may be offered in explanation of this tendency. First of all, the new working practices were often seen as an improvement, conferring a variety of benefits on the individual employee as well as on the firm. Even quite significant changes in working patterns were therefore considered acceptable, and their importance perhaps was overlooked.

Secondly, perhaps as a result of the perceived benefits, people on the whole adapted to the new working practices very rapidly, so that even quite traumatic periods of transition were rapidly forgotten once the new system was running smoothly. One supervisor who had been responsible for the installation of no fewer than 30 computerised systems at different branches of her office equipment leasing company in the space of just 18 months told us that employee reaction to the deployment of new technology and the associated changes in working practices fell into a predictable four-phase pattern:

Phase 1: total confusion for all concerned;
Phase 2: a growing conviction that the new system would never work;
Phase 3: the beginning of the belief that it might just work;
Phase 4: the realisation that it is now working.

Once phase 4 was reached people became familiar with the new system of working, and the traumas of phases 1–3 were quickly forgotten.

The third reason why changes in working practices were played down is that frequently they took place *before* the deployment of the technology, and were thus correctly perceived as not being the result of its use. Many of the changes in working practices originated in the need to reduce costs, a need which for most companies was given a sharper edge by the 1980–82 recession. In the intensified competition heralded by that recession many companies saw new technology as simply another weapon in the fight for survival, the principal weapons in their armoury being changes in working practices and changes in the attitudes of both management and workforce. In one engineering company the installation of a brand new machine shop was the culmination of a three-year exercise in improving communications and in breaking down entrenched attitudes held by both management and workers. In this company, the changes in both attitudes and working practices preceded the new technology. Although considerable changes in working practices took place, these were not seen as being the consequence of using new technology, but merely as part of a modernisation process which had begun several years earlier. The overall impression from our survey was that changes in working practices frequently helped to facilitate the deployment of new technology, rather than vice versa.

(c) Centralisation and Decentralisation

Historically speaking, the effect of technology on the centralisation of production and employment has been like a pendulum, in one era swinging towards increasing concentration, in a subsequent era swinging in the other direction. Thus the development of steam power in the early stages of the industrial revolution led to a concentration of production and employment in factories, and an associated growth of population in large towns and cities. Later, the replacement of steam power by electricity meant that factories could disperse.

In the early stages of computerisation of large firms, traditional data processing has tended to reinforce the existing dominance of the centre to the detriment of responsibilities at the periphery. Today, new technology makes possible decentralisation and local autonomy. In a few cases, such as electronic reservation systems or electronic currency, the new technology even confers positive advantages on the smaller local unit. In most cases, it is simply a question of management or organisational choice.

For example, in manufacturing the new technology provides for the possibility of greater autonomy in the workshop: a computer regulates many functions without referring constantly to management. However, equally, it may constitute a powerful means of centralisation: the network makes it easier to collect basic data and to follow up production operations in real time. In that case, the workers lose the limited degree of freedom they experienced under intermittent supervision, and may become even more integrated into the production process.

The finance sector provides some interesting evidence on the effects of new technology on centralisation. Banking, insurance and other finance companies usually have a common organisational structure, namely a head office with a regional branch network. The effects of new technology, when it is introduced, can be seen in the way in which responsibility shifts between the branches and head office. In principle, two opposing scenarios are readily conceivable. In one case, adoption of the new technology may tend to reinforce the tendency towards control from the centre of the organisation with a correspondingly reduced role for the peripheral elements (the branches). Alternatively, since new technology means that information can be accessed instantly from any convenient location in the network, there is the prospect of extreme decentralisation, with the re-emergence of a 'cottage industry' approach to information management, where each individual is able to operate independently from home.

That these choices are not just theoretical can be illustrated from the following cases of two finance companies with similar customers and range of services, but of rather different sizes. Both companies installed computerised customer accounts systems at their head offices, using similar technology, allowing them more rapid and more precise control over the movements in their customer accounts, notably in the important area of arrears and defaulting customers. In one company, the centrally-located computer provided each branch with arrears details for that branch's own customers. Responsibility for dealing with arrears was shifted from the head office to the branch, whose personnel, it was argued, had the local expertise necessary to deal with each individual case. The second company also used the new system to spot potential defaulters quickly with its central computer, but these were then dealt with by a series of letters from head office. In this company, both the autonomy and the responsibility of the branches were reduced, their only part in the debt-collection process arising when the default became so serious that a personal call on the customer had to be made.

New technology enhances the ability to control the flow of information in an organisation, and gives increased power to those who control this flow. Whether this tendency will lead towards greater centralisation or greater decentralisation will depend on the individual organisation and those who control it. They will have to balance considerations of the distribution of power within the organisation with considerations of profit. Even within one organisation, the continuing changes in technology may lead to cyclical fluctuations in centralisation and decentralisation trends over time. This should not surprise us. The development of semiconductor technology in the post-war period alone has, at various stages, first conferred, and then removed, cost advantages on different manufacturing technologies. The same developments have at some times favoured large companies, and at other times small companies. It should therefore not be surprising to discover that developments in new technology at times, and perhaps intermittently, favour decentralisation of organisation amongst some users of this technology, and at other times centralisation. The one thing which can be stated with confidence is that such continuing fluctuations only underline the disturbing character of rapid technical change in general, and of the rapidly changing nature of the new technology in particular.

Examples can always be found to support predictions of change in any particular direction. For example, it seems convincing to argue, in the case of consumer credit companies, that computerisation allows local agents to be by-passed entirely because it permits customers (motor car dealers) not only to use microcomputers which are programmed to calculate financial details and even to print the completed credit document ready for signing, but the same computers can also be linked directly to the mainframe of the finance house's head office. Thus deals may be centrally approved and payments undertaken by direct credit transfer between dealer and finance house, by-passing the routine activity of the branches.

On the other hand, some insurance companies expect to take advantage of the opportunities offered by computerisation to entrust one local agent with the entire handling of a set of client contracts. Previously, these companies were organised by type of disaster: one department specialising in automobile insurance, and another in fire insurance. Here, computerisation offers the opportunity of rendering redundant most of the head office's routine activities.

Although the development of semiconductor technology has led to a dramatic improvement in the reliability and technical performance of equipment, a highly centralised system of organisation is vulnerable to complete breakdown, should there be a failure in one particular part. There is no corresponding drawback to decentralised systems.

Where considerations of profitability do not indicate strongly either a centralised or a decentralised system of organisation, then there is scope for choice. In his famous report, *The Computerisation of Society*, Nora argues passionately that governments should discriminate positively in favour of decentralised systems and organisations. His motives are openly political. He believes that the pressure in the direction of 'structured, centralising networks' is so strong that it is necessary to counteract it, and that decentralisation is the only way of maintaining some degree of autonomy and responsibility for what he calls 'the weakest actors on the social stage'.[1]

EFFECTS OF NEW TECHNOLOGY ON THE ORGANISATION OF INDUSTRIES

The subject of this section is the effect of new technology on the size, distribution and functional specialisation of firms within an industry.

In all those markets in which it has been introduced, new technology has led to the faster provision of more accurate, more detailed and more comprehensive information than existed previously. This has had two broad consequences: intensified competition within existing markets, and the territorial extension of markets. In an increasing number of industries the effective market is now worldwide.

By improving the flow of information, new technology improves market performance. Specifically, it diminishes the possibility of profits being made by some market participants exploiting the relative ignorance of others. In other words, it limits the opportunities for profits being made by unproductive middlemen or through insider trading. This is particularly true for markets which are organised in 'hour-glass' form, i.e. where a limited number of operators act as go-betweens for scattered sellers and numerous buyers. An example is the fresh-food market in France, which has now been equipped with a computerised information network to diminish the disadvantages of both producers and consumers who are distant from the market and badly organised.

A number of other industries illustrate the growing globalisation of markets. Let us begin with the telecommunications industry. The telecommunications system forms a large part of the domestic capital stock of every nation in the world. For a long time the industry has been regarded as a natural monopoly, and is therefore either publicly owned or publicly regulated. Of all industries, apart from the electronics industry itself, telecommunications is the one which has been most dramatically affected by new technology. Indeed, the conjunction of developments in microprocessor technology with developments in communications systems technology has given rise to a new subject: telematics or information technology. The major recent developments in communications technology include the use of satellites, optical fibres and cable television to challenge conventional telephone landlines. The consequences for the telecommunications industry in the United States have been illuminating.

Until recently, a private company, AT&T, had a monopoly of all US internal telecommunications via landlines and digital exchanges. This company not only owned and operated the transmission system, it also had an equipment manufacturing subsidiary, Western Electric, and a large research and development operation, Bell Telephone Laboratories, where the transistor was invented in 1947. Being closely regulated by the federal government, the company's policy objective was not profit-maximisation, but sales-maximisation, subject to minimum profit constraints. Thus, they cross-subsidised unprofitable local transmission systems used largely by households with profits from long-distance calls, made largely by businesses.

When the communications revolution hit the United States, satellite communications companies began to offer cheaper long distance services to those who were heavy users of telecommunications, notably banks, stockbrokers, insurance companies and broadcasting companies. Other manufacturing companies grew up which were able to sell cheaper and more effective equipment for their own internal telephone systems, using the new technology. Finally, particularly large users of telecommunications acquired the option to build or use their own transmission lines or satellite systems (e.g. Citibank and Merrill Lynch).

In Europe, radio and television broadcasting has hitherto in most countries been a national monopoly or oligopoly closely regulated by the state. These limited sources of supply have already been augmented by the availability of such new technologies as video-cassettes, teletext, and cable and satellite broadcasting. New technology has also enabled the supply of individual programmes to national networks by independent producers, and this source of supply will be further enhanced by the advent of subscription broadcasting within the next decade—the decoding of encrypted signals by an individual receiver.

These developments illustrate the truth of Schumpeter's observation that all monopolies are temporary, and that even apparently 'natural' monoplies can be blown away by the forces of technical progress. Faced with competition forcing down the rates on their long distance calls, AT&T could no longer afford to subsidise local calls. Realising that it was impossible to have fair competition between one communications system which was regulated and others which were not, the federal government took the decision to go for total deregulation of the industry. The local telephone operating companies and the equipment manufacturing business were hived off from AT&T, which was left with its long distance lines and its laboratories.

One result of this reorganisation is that prices in the industry will now reflect costs much more accurately. In particular, the costs of the major business users of telecommunications have fallen significantly. Although there is no direct competition between telecommunications systems in different countries, it is unlikely that other countries will for long tolerate a situation in which their own businessmen competing with US businessmen in world markets should have to put up with domestic state-protected monopolies which cost more and give a poorer service than that enjoyed by their competitors.

A more direct threat of competition from US firms has forced the reorganisation of the finance industry in Britain. This is a direct consequence of the fact that new technology developments have made the market for financial services a global one. If the British had not reorganised their industry, they would simply have lost out in the global competition to financial service companies from the United States and possibly other countries.

Prior to the advent of the new technology and the development of a global market in financial services, the financial sector in Britain, and in many European countries, had been characterised by a large number of relatively small firms, each performing a single specialised function, e.g. retail banks, wholesale banks, insurance companies, stockbroking companies, building societies, etc. As a result of reorganisation, what is emerging is an industry characterised by a small number of relatively larger firms, each offering a range of financial services. Thus despecialisation has been the ultimate result of the introduction of new technology in this industry.

Despecialisation is also expected to occur in other industries, for example in health care. There, it is expected that computerisation will break up specialisms and extend the range of functions performed by the general practitioner. For example, he will be able to interpret an electrocardiogram, partly replacing the cardiologist. At the same time, many of the functions now performed by the general practitioner or by other specialists, such as anaesthetists, will in the future be capable of being performed by nurses or other paramedical assistants.

A final example of the way in which new technology has affected the organisation of an industry comes from newspaper publishing. Direct input by journalists and advertisement staff to a central typesetting computer is now the accepted method of production for many newspapers in the United States and in Holland. Newspapers in a number of other industrialised countries are either experimenting with single keyboarding or are in the final stages of securing a single keyboarding agreement with their unions.

The newspaper industry in Britain has changed more slowly than most. It can be regarded as two industries: the provincial newspaper industry and the newspapers published from Fleet Street in London, many with nationwide sales of several million daily. The provincial industry has faced increasing competition from a host of rival media in recent years: free sheets, regional radio and television. Circulation and advertising revenues have shrunk, and the rate of newspaper closures has accelerated. Those newspapers which have survived have done so by adopting new technology which has allowed them to cut their printing costs by around 30 per cent.

The Fleet Street newspapers had for long been notorious for their overmanning and other restrictive practices enforced by powerful trade unions. These costs provided a formidable barrier to the entry of any new competitors. The advent of new technology has made it possible to produce and circulate a more attractive and up-to-date newspaper, with colour printing, at a fraction of the traditional cost. The result is that, for the first time for 50 years, two new titles have appeared, published by very much smaller firms. Some of the existing newspaper publishing houses have responded by deploying the new technology, and cutting their production staff by up to 70 per cent, but others will certainly go bankrupt. Thus the effect of new technology has been to lower the barriers to entry, and altar the size distribution of producers in the industry.

NOTE

1. It is ironic that a proposal for decentralisation should emanate from the centre. There is no evidence that a contemporary policy of decentralisation is any more popular with the inhabitants of the regions of France than was the policy of centralisation ushered in by the French Revolution. Indeed, there is some evidence to the contrary: General de Gaulle resigned following the defeat of his proposal for regional decentralisation in a nationwide referendum.

5 Effects on Trade Unions and Government Administration

This chapter discusses the effects which the deployment of new technology has had on the attitudes and organisation of trade unions, and on the attitudes and organisation of staff in government administration.

The response of trade union leaders throughout the industrialised world to the advent of new technology has been to issue warnings, to hold conferences and to attempt to formulate a policy. Policy papers have addressed such questions as:

- Is the new technology fundamentally different from that of the past?
- Is the rate and diffusion of technical change increasing?
- Can the institutions of a market economy cope with change, while ensuring a return to full employment?
- How are the costs and benefits to be distributed?
- Who is to control the new technology?
- How will new technology affect the organisation of work?

From the many attempts to answer such questions, Corina (1983) has distilled what might be termed a composite trade union leadership view, which can be summarised in the following seven propositions:

1. Blanket opposition to new technology is neither practical nor desirable.
2. New technology is likely to produce a loss of jobs in the short run, and perhaps even in the long run.

3. However, new technology is not primarily to blame for the recent increases in unemployment in the industrialised countries.
4. Unions on the whole do not distinguish between the concept of preserving jobs (the particular set of tasks performed by a worker) and the concept of preserving employment.
5. A shrinking volume of available work should be translated into a shorter working time for more people. Reductions in the working week, longer holidays, flexi-time and early retirement all remain trade union objectives.
6. Governments should attempt to discourage the labour-displacing effect of new technology.
7. Workers should be protected against deskilling, and against technology which could dehumanise work, (e.g. by electronic surveillance) or which could adversely affect health and safety.

These are the views of the trade union leadership, but it is well known that the rhetoric of official trade union pronouncements are no guarantee of corresponding workplace attitudes. As we shall see, this is as true of new technology as it is of other challenges which face the unions. In this section, we deal with the trade union response at the workplace level to the introduction of new technology. Further discussion of the effects of new technology on the trade unions at national level is contained in Chapters 8 and 10.

THE DILEMMA OF THE RECESSION

It is impossible to divorce any discussion of the impact of new technology on industrial relations from the economic background against which deployment occurs. Most of the new technology in our survey was installed during the recession of 1980–82, and the implications of the economic circumstances of that period must clearly be understood. In a recession, both management and unions are faced with dilemmas. For management, recession makes more urgent the need to invest in new technology in order to remain competitive. At the same time, the recession means that it is harder to earn the revenues required to finance the new investment.

Trade unions face a dilemma of their own. They are concerned about the possible impact of new technology on their members' jobs, which are particularly at risk during a recession. At the same time, they recognise that long-run job security for their members can only be achieved in competitive firms, and that to remain competitive firms must invest in new technology. Unions are faced with the knowledge that if they consistently set their faces against change, they are likely to be overtaken by events. Not only will change eventually occur against their wishes, but more adaptable unions are likely to gain at the expense of those which oppose change. One of the effects of a recession is of course to drive out of business those firms which, whether because of the attitudes of their management or their workforce or both, are the least adaptable. In this sense it is possible to speak of a 'remedial' recession.

Once the decision to adopt new technology has been made, the dilemma is essentially resolved for management. The recession serves as an impetus to ensure the success of the deployment: it *has* to work because the firm cannot afford failure. This in turn may strengthen the determination of management to effect necessary changes in the face of union opposition. At the same time, the implementation decision exacerbates the dilemma of the union, thus creating a potential for conflict.

In fact, we found very little evidence of such conflict in the companies we surveyed. On several occasions we found that shop stewards had actively assisted in the deployment of new technology, and had contributed to its smooth running once the need for it had been clearly explained. During an interview with two officials of Britain's second largest union, the Amalgamated Engineering Union, they said that their union had often been instrumental in encouraging laggard companies to adopt new technology in order to remain competitive. The potential for conflict posed by the dilemmas of the recession only infrequently developed into actual disputes, the vast majority of deployments taking place with either the tacit or active support of the unions involved. The reasons for this were threefold:

1. Trade unions in the private sector are rarely hostile to new technology in principle, even if they have reservations about its use in particular instances.
2. Partly as a result of the recession, managements have become more determined to implement their decisions on the adoption of new technology even where there is some evidence of union opposition. Although our interviews revealed little evidence of the recent recession resulting in a more arbitrary or authoritarian management style, both management and workers were of the firm opinion that the 'pendulum of power' had swung in the managers' favour since 1979. More than one shop steward expressed surprise at the hard line that management were prepared to take on new technology, while the growing tendency for management to disseminate information directly to the workforce has done much to diminish the importance of the unions in the implementation process.
3. The overwhelmingly positive reaction of the great majority of workers to new technology would make it very difficult for unions actively to oppose its introduction even if they so desired. Union members, and even shop stewards, expressed to us the view that unions had been unduly resistant to change in the past, even when clear benefits had been established. Most workers we interviewed took a dim view of anyone standing in the way of the increased personal marketability which training in new technology could bring, even if that meant demolishing the traditional demarcation within and between craft organisations on which so many unions are built. The union officials' doubts about their ability to carry their members with them in a dispute inevitably weakens their bargaining position.

These findings are largely in accord with the results of other studies of trade unions and technical change. In a comprehensive review of the situation in Britain, Northcott, Fogarty and Trevor (1985) wrote:

> unions power to obstruct management decisions was in fact much more limited that it often appeared, not least to managers themselves ... the typical situation...was that unions might argue over the detailed implementation of management's strategic decisions but, if faced by determined management, had little control over the decisions themselves. (p. 21)

And later:

> if unions had more ultimate power to obstruct change and adaptation, it is not at all clear that they would generally have wished to use it, or that their members and shop stewards would have supported them in doing so. (p. 22)

A similar study comparing the deployment of new technology in Britain with France and Germany found, somewhat surprisingly, that shopfloor trade union opposition to new technology was less intense in Britain than it was in Germany and France. A survey of more than 3800 European factories found that the management in only 7 per cent of the British factories classified trade union opposition at shopfloor level to the introduction of new technology as being a 'very important difficulty', whereas the comparable figure for Germany and France was 14 per cent. This is still a fairly low figure, even for France and Germany.

It is possible therefore to characterise the general reaction of trade unions to the proposed deployment of new technology as being one of qualified approval ('yes, if'), rather than negotiated rejection ('no, unless'). Qualifications to this approval arise under three main headings: (1) no job losses, (2) more pay, and (3) assurances concerning health and working conditions.

(1) No Job Losses

While the majority of unions accept that new technology helps to protect the long-term employment of their members, they are deeply concerned about the short-term impact on jobs. Almost half the companies which we surveyed gave guarantees of no compulsory redundancies at the time of implementation; in the printing industry this guarantee was extended to no job losses whatsoever. In other words, even jobs which became vacant through natural wastage would be filled.

It was notable that the incidence of no-redundancy

guarantees was not always associated with the presence of strong trade unions. Five of the firms in the finance sector gave such guarantees, despite the traditionally low level of unionisation in this sector, while only two such assurances were given in the highly organised engineering industry.

(2) More Pay

Unions frequently attempt to obtain increased rates of pay for operators directly affected by new technology. Outside the printing industry, such claims proved totally ineffective. The general tendency was for management to resist strongly any attempts to secure increased pay for the use of new technology, though in some cases guarantees of no reduction in earnings were given.

A hard line on pay taken by management did on occasion cause resentment among shop stewards; they felt that since one of the effects of new technology was a marked increase in labour productivity, and because their members had to adjust to new working patterns, more cash should have been offered. The general management view was that operators would benefit from the retraining which they had received, and that any increase in wages would only reduce one of the principal advantages of new technology, namely lower labour costs. With few exceptions, the management view prevailed, and the unions retired empty-handed. One of the contributory factors to this outcome was the willingness of union members to forego wage increases in order to achieve what they saw as the enhanced status associated with the operation of new technology.

(3) Health and Safety

There was a third area of concern for the unions: the health and safety aspects of new technology. For example, concern was expressed both before and after deployment about the possible impact of VDUs on the health of regular users, a debate which remains inconclusive. To our surprise, we found that many workers preferred their union to represent their views to management on these issues rather than on such higher-level issues as jobs and pay. This is a point to which we return in the following section.

DIFFERENTIAL UNION REACTIONS

Priorities frequently differ between unions, so that a united front on new technology in a multi-union plant is seldom achieved. Examples from Britain include the wrangles between AEU and TASS about whose members should be responsible for the editing of CNC programmes, and the conflict between the print unions SOGAT and NGA and the EEPTU and AEU over the use of new technology at News International's plants in London and Glasgow. Such inter-union disputes serve to weaken the bargaining position of the unions, especially in multi-union plants and offices. Despite the TUC's campaign for the establishment of new technology agreements, we encountered only two such agreements in our survey. Indeed, even the TUC's own video 'Technology at Work' was unable to come up with a single example of a successful new technology agreement in a private sector firm. Despite the TUC's expressed desire for widespread, early and active involvement of unions in the deployment process, our finding was that unions mostly played only a marginal role in this process, being by-passed by the management's preference to deal directly with the workforce.

WORKERS' PERCEPTIONS OF UNIONS

Some clues as to why unions appear to play such a marginal role in the implementation of new technology are to be found in workers' perceptions of trade unions and of the unions' attitude to new technology.

Most factory workers we interviewed were favourably inclined towards unions in principle, a view which was almost as common among non-unionised firms as those in which there was an active membership. However, the role which their members expected unions to fulfil was clearly not that of a full partner in economic and social affairs enjoyed by the TUC under previous Labour administrations. Both shop stewards and ordinary members felt that the unions' efforts should be directed more to local workplace issues: for most union members, improved canteen facilities were more important than a new social contract. A view which was consistently expressed to us was that unions were a necessary check on the potential exercise of arbitrary authority by management. 'Useful, but too political' was one view which summed up the general attitude.

In companies where there was no official union recognition, workers tended to assert that there was no need for them because they already enjoyed satisfactory pay and conditions. Indeed, in the electronics industry there was even an attitude that intervention by unions could prove harmful to existing pay and conditions. This did not prevent those who expressed such views holding favourable opinions about the desirability of unions in principle, they simply felt that there was little need for the presence of unions in their particular firm.

Those union members who claimed to be aware of their own union's policy towards new technology were frequently wide of the mark. Unions were seen as being much more hostile to new technology than was in fact the case. A typical reaction from a member of BIFU (the financial sector trade union) was: 'they start off by giving a flat "no" to new technology, then negotiate'. As we have seen, this was quite untrue.

It may be that people have a greater capacity for recalling areas of dispute than areas of agreement. One piece of new technology which led to disruption will be remembered long after the much larger number whose implementation passed off without incident. However, if union members can have such a misconception of the attitudes of their own organisation, it may not be surprising that trade unions have so often acquired a public image of being obstructive and resistant to change.

We can sum up this discussion on unions and new technology by saying that, while unions are approved by most workers as an essential insurance policy against management misbehaviour, not many workers wish their union to come between them and the management when it comes to negotiations about new technology. The question of how the unions should respond to these attitudes, and to other changes brought about by new technology, is discussed in Chapter 10 below. Meanwhile, we can summarise the role of unions as perceived by their members, so far as the deployment of new technology is concerned, as being necessary but marginal.

THE EFFECT OF NEW TECHNOLOGY ON GOVERNMENT ADMINISTRATION

Ten years ago, the agencies and departments of central and local government along with many other public sector agencies in the industrialised world seemed ripe for computerisation. Like banks and finance companies in the private sector, they employed a large clerical staff performing routine office functions, and therefore the scope for productivity gains seemed large.

But office computerisation has not proceeded as fast in the private sector as had been expected ten years ago. This does not mean to say that it won't happen—it just hasn't happened yet. Some developments have taken off, notably word-processing, while others which are well within the scope of existing technology, such as video-conferencing, have not. And the rate of deployment seems to have been even slower in the public sector than it has been in the private sector. Why?

The pressures to introduce new technology come from quite different sources in government administration than they do in organisations which are part of the market economy. In government departments and agencies, there is no competitive cost-cutting, which requires innovation as the price of survival. There is no question of 'automate or liquidate'—the stark alternatives which can sometimes face organisations in the competitive commercial world. It may be asked, in the absence of any motivating forces why should there be any change at all? It will be said that where there are evident savings to be made, i.e. where potential increases in productivity are available, then political pressures directed to budget limitation if not cutting, combined with good departmental management practices, will ensure the appropriate rate and manner of deployment of new technology in government administration. The trouble is that there are strong organisational resistances.

First of all, the existence of measurable productivity improvements resulting from office computerisation, by itself, is very hard to prove. Those who have tried have sooner or later found themselves embroiled in the fact that subjective factors are the ultimately important ones in any evaluation. Because it is so difficult to demonstrate the value of office automation, one American salesman only seeks out potential customers who don't need to be convinced, and then persuades them that he has a better product to offer.

The fact is that the really large increases in productivity which are available through the computerisation of government administration come about in much the same way as do the savings in the administrative sections of commercial organisations, namely via reorganisation of tasks to eliminate redundant, overlapping or ineffective functions. In government as in the private business sector, the deployment of new technology often rides on the back of reorganisation. The difference is that, in business office management, new technology often follows or accompanies reorganisation, whereas in government, the reorganisation of working practices often follows the introduction of new technology. However, in government administration, reorganisation can encounter much more powerful resistance from staff than is the case in the private sector. Examples from three different countries are instructive.

Immel (1985) tells how in the late 1970s, Edward Scott Jr found himself responsible for 20,000 people and a budget of $17 million as Assistant Secretary of the United States Department of Transportation. Such was his frustration with the mountains of paperwork which he encountered that he got a special appropriation from Congress to build his own tailor-made, in-house computerised office system. It was run by a mini-computer linked to terminals on the desks of managers. But Scott eventually left government service to set up his own company which supplies automated office systems including such functions as word processing, electronic filing and database access. He now feels that it will take another ten years for the technology such as the system he had developed to be accepted in most offices, although he has no doubt that it will happen eventually. His confidence, however, is based on faith rather than on hard numbers.

The progress of new technology in the French government is described by Nora (1978). The computerisation of government administration began in the early 1970s, and proceeded in a piecemeal fashion with each department being left to take its own initiatives. The spread has been very rapid but also very uneven: some departments like that of tax collection have developed their own data-processing empire, whereas others, like the Ministry of Justice, have done little or nothing. In the case of really large-scale projects such as the computerised procedures for customs clearing at major airports, some have taken nine years from the design of the system to actual operation. Generally speaking, computerisation in French government is more advanced in the older and richer departments such as the Ministry of Finance, the armed forces and the police than it is in such other departments as Education, Health or Justice or in local government.

This lack of coordinated development has meant wasteful duplication and incompatibility. At the boundaries of two different computer systems, even within the same department, there has to be manual retrieval of data. The classic example occurred when two different departments of government each established their own land database: one for tax purposes, the other for development purposes. As a result, 'data processing ... [has made] it more difficult to set up a land administration in the future, when it might be desirable to do so'. Each department or each branch within a department having made its own isolated investments, it then resisted any attempt at coordination, which might be seen as reducing its own influence within the administration.

In England, the attempt to deploy new technology in a number of branches of local government was the apparent cause of a rash of strikes which affected more than eight local authorities in the course of the year 1984. A remarkable fact was that all the authorities concerned were controlled by the Labour Party, which was politically sympathetic to the aspirations of the trade union (NALGO) which represented the majority of the local authority staff concerned. In all cases, 'no-redundancy' agreements had been offered, but even those councils which offered genuine consultation about the proposed reorganisation ran into a 'brick wall' of hostility. Although closer examination of these incidents suggests that the principal factor at work was staff opposition to organisational change, rather than to new technology *per se*, it should be remembered that the greatest benefits of the latter cannot be usually obtained without the former. The lesson of these events would appear to be that the willingness of government staff to resist change should not be underestimated. This is a very different experience from that which was encountered in the private business sector.

6 Effects of New Technology on Numbers of Jobs

One of the aspects of new technology which has attracted the greatest attention is its likely effect on employment. During the late 1970s and early 1980s, a number of studies were published, some from usually authoritative sources, forecasting dramatic job losses in the industrialised countries as a consequence of the imminent deployment of new technology. Some of these forecasts were quite apocalyptic in scale, and achieved a certain notoriety on that account. Few commentators today believe that new technology is primarily responsible for the rise in unemployment which has taken place throughout the industrialised world in the 1980s, or, to the extent that it is, not in the way suggested by the earlier forecasts. This chapter discusses the relationship between new technology, new jobs and unemployment. The effects of new technology on skills, and other human effects, are discussed in Chapter 7.

It is first of all necessary to make a distinction between the *direct* displacement of jobs, i.e. jobs lost as a result of the introduction of new technology at the place of work, and *indirect* displacement, for example loss of jobs in one plant due to successful cost-cutting by a competing plant which has adopted new technology. There are also corresponding direct and indirect job gains. Most of those who have studied the employment effects of new technology have come to realise that the *indirect* effects (both positive and negative) are likely to be more important than the *direct* effects in determining the net effect of new technology upon the total amount of employment in any economic system. We begin however with the purely direct effects, with which most quantitative studies have hitherto been largely concerned.

DIRECT JOB LOSSES AND GAINS

In his capacity as Head of the World Centre for Computer Sciences and Human Resources, Jean-Jacques Servan-Schreiber forecast in 1982 that by the year 1990 50 million people in the industrialised countries would be unemployed because of new technology. In Germany a study by the Siemens electrical equipment manufacturing company estimated that by 1985 over 40 per cent of all clerical jobs in Germany would disappear as a result of the use of new technology. The Nora report (1978) anticipated that 2½ million office jobs would be lost in France by the year 2000.

For the United States, the influential magazine *Business Week* (1981) estimated that 45 million jobs could be lost as a result of new technology—38 million in offices and 7 million in factories. For the United Kingdom, Barron and Curnow (1979) predicted a 16 per cent loss of jobs over the following 15 years. In a study published in the same year, Jenkins and Sherman estimated that 1 million jobs might be lost by 1983, 3.8 million by 1993 and over 5 million by the year 2003. A 1979 study by a clerical trade union (APEX) predicted a loss of ¼ million office jobs in Britain by 1983, while another white-collar trade union (ASTMS, 1979) suggested a loss of 2.6 million office jobs by 1985.

All of these estimates of net job losses appear in retrospect to have been greatly exaggerated. There are two broad reasons why they have proved to be so wrong. First of all, many of them are based on an inappropriate generalisation from the results of particular case studies to the economy as a whole. For example, various studies have been made showing the remarkable gains in productivity in offices made possible by the use of word processors. Crudely applying possible or actual displacement rates based on particular studies to wider groups of workers overlooks two considerations. In the case of clerical work, huge productivity increases can only be realised in large centralised office systems which are the exception rather than the rule. Secondly, when higher labour productivity is achieved, it is often used for purposes other than job displacement. In offices, new technology can be used to generate or to communicate more information or new kinds of information, or to achieve a higher quality of presentation. In fact some businessmen go so far as to argue that office automation will not destroy jobs at all, believing that when computerisation is properly applied, new job opportunities are created. In this view, staff can be released from routine and repetitive tasks, and switched to more effective work—for example, on sales promotion—leading to an overall growth in the firm's business. In manufacturing, new technology can likewise be used to realise a greater volume of output or an improvement in the quality of the product. When these things happen, and employment is maintained, the result is sometimes described as 'jobless growth'.

Let us look at the evidence about what has actually happened in terms of direct job losses. While there have been a number of case studies at the micro level, the most comprehensive evidence comes from a survey carried out by three European research institutes of more then 800 factories in France, Germany and Britain (Northcott, Knetsch and de Lestapis, 1985). This survey found that three-quarters of the factories using new technology reported no change in the numbers employed as a result. In the remaining quarter, some even had increases in employment, although twice as many had decreases.

The overall net loss in employment experienced by the factories surveyed turned out to be only one or two jobs per factory. The total net direct loss in jobs as a result of new technology has therefore been estimated at somewhere in the region of 15,000 a year in Britain and in Germany, and about half that number in France. This rate of job losses represents less than 0.5 per cent of total employment in manufacturing in all three countries, and less than 5 per cent of the loss of jobs in manufacturing from all other causes. It should, of course, be borne in mind that such modest rates cannot necessarily be projected into the future. More advanced technology, together with wider diffusion, could have greater direct effects on employment than has been noticed so far.

INDIRECT JOB LOSSES AND GAINS

It is generally recognised that the major part of the influence of new technology upon jobs will not arise from direct labour displacement, but rather indirectly through changes in labour requirements elsewhere in the economy. The other main reason for the failure of extremely pessimistic job forecasts to be realised is that they failed to take into account the number of new jobs that would indirectly be created by the deployment of new technology.

On the negative side, a lower level of real incomes resulting from the direct displacement of labour will have multiplier effects. However, these are mainly likely to be local, and not nearly as important as the effects of competition. Those firms which deploy new technology may gain market share at the expense of firms using existing technology, and rival firms which unsuccessfully deploy new technology. Loss of market share often means loss of employment. The firm losing market share cannot distinguish the source of its rival's advantage, and therefore it is impossible to make conclusive quantitative estimates of the size of such indirect effects. Some examples, however, are provided by the experience of the changeover from electro-mechanical to electronic cash registers, and the experience of the Swiss watch industry. The direct displacement of jobs by new process technology in conventional watch production has probably been insignificant. But the indirect effect of employment in this industry was dramatic.

Digital watches manufactured in Japan and the Far East, and incorporating new technology, swept conventional watches from world markets, almost destroying the Swiss watch industry in the process. By 1982, Ausag, the largest firm in that industry, had shed 2200 jobs, and both it and the second largest firm were on the verge of bankruptcy. In the same year, exports of Swiss watches fell by 40 per cent. The industry was only saved by the introduction of a new product, a plastic battery-powered watch embodying microprocessor technology, entirely different from watches traditionally produced in Switzerland.

Indirect job gains are likewise created elsewhere in the

economy by the multiplier effects of any initial increase in profits and other incomes arising from the direct deployment of new technology. There will be other positive indirect effects on employment. First of all, there is the employment required to produce the new equipment which will embody the new technology. And secondly, there will be employment created by the need to produce the components and services required for the new equipment. Finally, increased exports may result from the improved price and non-price competitiveness of goods embodying the new technology.

Taking together direct and indirect gains and direct and indirect losses, what will the net effect of new technology on employment be? It is impossible to give a general answer to this question: it depends on the particular circumstances. The effect on employment in any one country is likely to depend amongst other things on (a) the extent to which the increased requirement of capital equipment is met from domestic sources, and (b) the ability of firms in that country deploying new technology to improve the non-price competitiveness of their products in export markets.

Some calculations carried out at Warwick University by Whitley and Wilson (1981) showed that the net effect *could* be positive in some circumstances. They estimated that 340,000 jobs might be lost in the UK between 1980 and 1990 as a result of the introduction of microprocessor technology, but that during the same period 420,000 jobs could be created from the same source. In their calculation, all industries which deployed new technology, except mechanical engineering, had lower levels of employment at the end of the period than they had at the beginning, but these job losses were more than offset by gains in other industries, notably in construction and services, which occurred mainly as a result of increases in consumer and investment spending associated with the introduction of new technology. What can be stated with confidence is that, so long as there is no limit to human wants, there will always exist the possibility of maintaining employment in the face of technical progress, no matter how dramatic are the increases in productivity associated with particular new technologies.

However, even when the total number of jobs created by new technology exceeds the total number which is lost, the introduction of new technology can still give rise to substantial unemployment. This is because it is not at all likely that the new jobs created will match those which are lost, in terms of time, place, industry or occupation.

MISMATCHES

There are mismatches in time, because there is no necessity that new jobs will become available *at the same time* as the old ones are lost. Likewise there can be *spatial* mismatches. With the free international movement of goods and capital, it is quite possible for the majority of the new jobs arising from some new technology to be created in one region (e.g. Japan or the USA), while the jobs which are displaced are displaced preponderantly in another region (e.g. Europe). From the point of view of a national community, part of the employment benefits from the deployment of new technology depends on the country of origin of the new capital equipment, and the terms on which it is supplied. For example, in 1983 only 34 per cent of all new industrial robots installed in the UK were produced in the UK.

Nor do the jobs created need to be in the same *industries* as those which disappear. Indeed, at any moment of time in the industrialised countries the picture is one of considerable differences in the relative growth rates of different industries, with some experiencing rapid expansion even throughout what is believed to be a general recession, while others are stagnant or even declining during periods of general prosperity. Even if new jobs were to become available in the same region, at the same time, and in the same industry in which the old ones were lost, there may still be problems of employment adjustment. For example, suppose a firm decides to computerise simultaneously its office and production operations, so that it makes redundant some middle-aged unskilled male labourers, while at the same time seeking to add to its skilled female word-processing staff, perhaps on a part-time basis. There is evidently a mismatch between these job characteristics which can lead to unemployment coexisting with unfilled vacancies.[1]

It is quite possible that, in this way, new technology is in part responsible for the present high levels of unemployment, because there is some evidence that the deployment of the new microprocessor technology does have an asymmetric effect on the demand for labour skills. Specifically, it decreases the demand for unskilled labour and increases the demand for skilled labour.[2] Such studies, however, relate only to the direct demand for labour. There is no reason why the introduction of new technology should not create an indirect demand for unskilled labour as, for example, in the leisure and catering service industries, security guards, hospital workers, etc. It is apparent to the casual observer that there is no shortage of unskilled work needing to be done. Why then should there be such an unfavourable unemployment/vacancy ratio for unskilled as compared to skilled workers? One answer to this question may be that demand and supply for unskilled workers could only be balanced by a widening wage differential relative to skilled workers. Such a widening might seem politically unacceptable to an egalitarian climate of opinion. For example, Braun and Macdonald (1978, p. 198) complain that the jobs destroyed by new technology are those for unskilled people and that the jobs created are often 'the really badly paid ones, such as in catering, or the really unpleasant ones, such as refuse disposal'. While it is true that many unskilled jobs are badly paid relative to skilled jobs, there equally cannot be much doubt that the real income of those engaged in unskilled jobs has risen in absolute terms in recent years.

Secondly, it may be noted that while some of the indirect demand for unskilled labour services is generated through the market sector of the economy, a large and unsatisfied demand comes from the public sector: health, street cleaning, security and social work, all of which are funded by revenues transferred from the taxation of activities in the market sector. Such employment opportunities are therefore limited by the rate of growth of the market sector of the economy, as well as by the apparent tendency of public sector employing authorities to expand skilled jobs and to reduce unskilled ones.

CONCLUSIONS

Thus we can see that new technology may be responsible for creating unemployment, not because it is a net destroyer of jobs, but because the introduction of new technology implies a change in the type of jobs to be performed in the economy. A change, of course, is never total: it only affects certain functions, leaving others unchanged. But it may make the functions on which particular occupations are founded redundant, and equally, by creating new types of functions, give rise to new occupations. It is evidently one of the tasks of government to ensure, so far as possible, a smooth transition from one occupation to another. This question is further dealt with in Chapter 9. In the meantime, it may be noted that in an advanced industrialised economy, the relationship between new technology and employment cannot be usefully discussed at an aggregate level. It is only when we disaggregate that the causes of unemployment, and its possible solutions, come into view.

The labour force which is available for employment in any country is not simply a stock, determined by demographic factors, but a flow, strongly influenced by socio-economic factors. Work and employment are by no means synonymous: there is a range of activities which can occupy individuals without employing them—education and housework being the most common of these. The relationship between remunerated work and the actual number of jobs is determined by the length of the working week. Historically, technical progress, and its related increase in productivity, has been associated with a long-run decline in the number of hours which constitutes the standard working week. There has also, over the same time period, been an increase in the rate of participation of adults of working age. New technology, through the increased productivity which it brings about, opens up the possibility of a reduction in the number of hours worked per week while keeping the real income of those in employment constant, and at the same time *increasing* the number of jobs available, albeit some of them on a part-time basis.

For example, in the United States, the number of part-time

employees as a proportion of total employment has increased from 14 per cent in 1968 to 17.4 per cent in 1985. Two-thirds of those part-time workers were female, and 27 per cent of women who were employed had part-time jobs, while only 10 per cent of men were employed part-time. Indeed, the advent of new technology may simply reinforce the trend towards increased part-time working, which has been noticeable in the industrialised countries for some time.

NOTES

1. This is only an illustrative example; it should not detract from the important point that when a redistribution of the labour force among activities takes place within one organistion, then, in general the opportunities for avoiding unemployment may be easier. See Chapter 8 below.
2. See Northcott, Fogarty and Trevor (1985), p. 144, Table 7.

7 Effects of New Technology on Skills

Concern about the impact of new technology on skills is by no means new; it was a topic of debate in the 1830s when Babbage analysed the effects on weavers of the introduction of mechanical looms. However, it was not until the development of 'scientific management' in the 1920s and the move towards large-scale automation of plants that concern became expressed more widely about the potential extent of deskilling and the associated lack of job satisfaction for large sections of the workforce. The first part of this chapter deals with the effects of new technology on skills, and the second part with other effects on humans at work.

Two opposing views of the impact of new technology on skills can be simply stated. The first is that technical change exhibits a long-term trend in the direction of making skills unnecessary and crafts redundant. According to Braverman (1974), the individual factory or office worker is eventually reduced to no more than an appendage to a machine, and exercises only a marginal influence on the process of production.

The alternative view is that new technology is a means of liberating the worker from the boredom and drudgery of much of his working life. By eliminating the repetitive and/or physically demanding elements of the job, the individual is free to pursue its more creative and fulfilling aspects; furthermore, his or her personal skills become enhanced as the sophistication of the technology increases.

The evidence which we have uncovered does not readily fit either of these two views. This may be for two principal reasons: first, because skill is a multidimensional phenomenon; and second, because it is a subjective phenomenon, the perception of which varies significantly from one individual worker to another. Rather than thinking of new technology as a factor which increases or decreases the overall level of skill inherent in any individual or task, it makes more sense to consider it as a means of rearranging the constituent elements which characterise the skill associated with that individual, reducing or eliminating some elements and introducing or enhancing others.

The elements shown in Figure 7.1 (a and b) are not meant to be exhaustive, but they are indicative of some of the dimensions which, taken together, constitute an individual's 'skill'. Thus, a skilled turner in the engineering industry who now operates a CNC machine may recognise a reduction in the craft element of his skills but perceives himself to be no less skilled overall because another important element, his marketability, has increased.

The perception of skill is essentially personal, and is closely bound up with status and self-esteem, both at work and outside work. Whether or not the net effect of a particular change in technology is 'deskilling' will therefore depend on the relative weights which the particular individual attaches to each element of his or her skill package. This approach may help us to explain an apparent paradox of skills. This refers to the possibility of perceiving an individual *worker* as being no less skilled, or even more skilled, through his or her use of new technology, while at the same time recognising that the task which that individual performs may be less skilled because of a reduced craft input. Such a paradox is often found in circumstances where an individual worker has served a traditional craft apprenticeship before moving to operate computer-controlled machinery. If we think of skills as being multidimensional, and of new technology as being a means of altering the constituent elements, then the paradox is resolved; the individual can be more skilled and the task less skilled at the same time.

A range of effects of new technology on their skills was perceived by the interviewees in our survey. They can be reviewed as follows, in order of decreasing satisfaction,

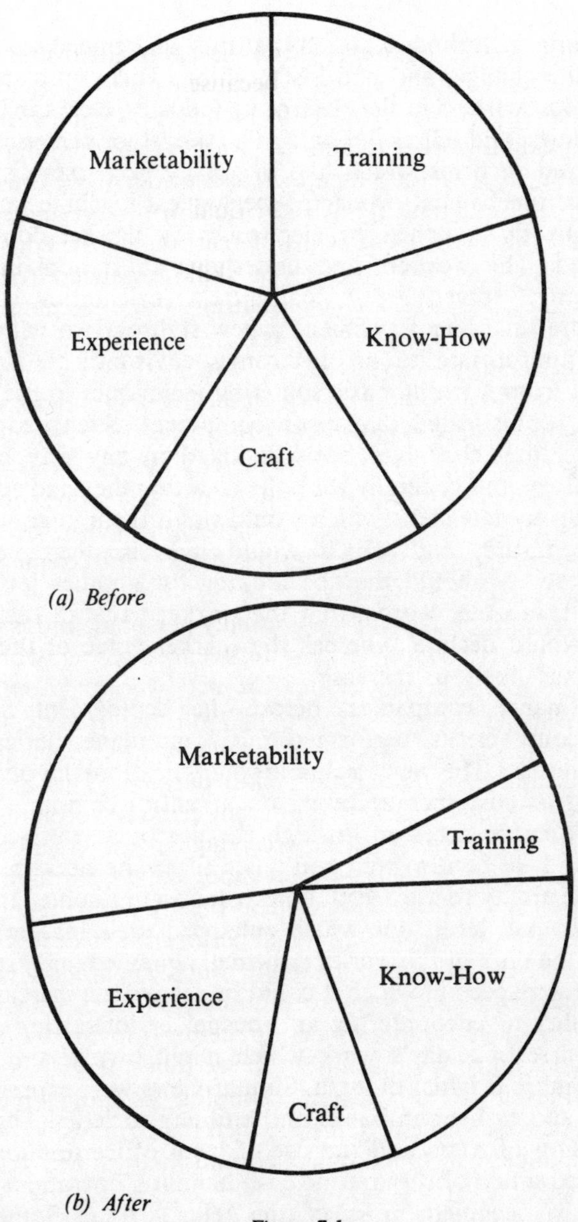

Figure 7.1:
Elements of an individual worker's skills (a) before new technology, and (b) after new technology

beginning with those who felt that they had gained something from the change, and had lost nothing. This group included shopfloor workers in the electronics industry, clerks in finance companies, and tellers in banks. The shopfloor personnel were employed in firms which had originally been established to produce mechanical or electro-mechanical machines and had subsequently switched to electronics as market conditions changed. The workers had undergone substantial in-house retraining, sometimes a complete reorientation of job structure, in order to achieve a new skill pattern which was more appropriate to an electronics environment. Training ranged from learning basic soldering techniques to the use of highly sophisticated diagnostic equipment. Such people did not feel that they had been deskilled in any way by new technology, but generally took the view that they had acquired more up-to-date skills which would stand them in good stead in the future, and which would increase their personal marketability should they be looking for another job. They realised, in other words, that the market value of their 'old' skills would decline, whereas the market value of their new skills was likely to increase.

In finance companies, before the deployment of new technology, clerks performed fairly mundane clerical and filing duties. The new technology applications involved the placing of customer accounts on a central computer which could then be accessed through the use of a VDU, replacing thousands of handwritten and typed files. The clerks reported that the ability to use a VDU represented an undoubted increase in their skill level, one which enhanced their marketability. While the bulk of the work performed remained fairly routine, the greater speed at which it could be carried out increased the possibility of encountering an unusual or interesting case in the course of a day's work which might involve the use of rather more original thought. Similar views were expressed by tellers and cashiers in banks and building societies. The rapid expansion of ATMs and the use of front office terminals has removed much of the routine cash handling operations which formerly occupied most of the teller's time. Some such transactions are still carried out, but the time saved by the removal of most of them has often been used creatively to turn the teller into a sales person and problem-solver. Many financial institutions, especially building societies, now expect their tellers to be aware of the full range of services offered, and to sell effectively to the customers with whom they come into contact. Many tellers expressed considerable satisfaction at the selling and explanatory aspects of the job, and clearly felt themselves to be of a higher status than the teller of 20 or even 10 years ago. In the cases of both the clerks and the tellers, the overwhelming weight of evidence was that new technology had not diminished either skill or job satisfaction from the tasks. Indeed, in a significant number of cases it had widened the range of activities which were carried out in the course of their working day and had enriched the nature of the job.

A second group of workers consisted of those who felt that in moving from conventional to computer-controlled equipment they had gained a valuable skill which more than compensated for the fact thay they no longer practised their traditional craft. Into this category fell operators of CNC machines, typesetters, compositors, and other engineering and printing workers. CNC machines are controlled by a microprocessor, and programmed to carry out a detailed sequence of machining operations. The operator will normally be a time-served craftsmen, such as a turner or grinder, with considerable experience of a relatively narrow range of tasks, on whose skill much of the quality of the finished product would have depended under the older conventional technology. The use of CNC equipment moves the responsibility for machining from the operator to the computer. The skills required to operate the new machines are minimal, and could be learned by anyone in two or three days. Yet the former craftsmen did not see themselves as being less skilled as a result of the move to CNC equipment. Why not? First, they claimed that they could return to their conventional machining skills at any time they were required to do so, and hence these skills were not lost to them. Secondly, they argued that while a complete novice could quickly be trained to operate CNC machinery, he would lack the skill and the experience to identify when a fault had occurred in the machining process. We found this a very common point of view in all skilled occupations, that while a new machine might remove the need for a craft skill to be practised, it did not remove the need for judgement about the quality of the finished product, judgement which could only be based upon traditional skill and experience. Lastly, the CNC operators made much of their ability to program the machines, or at least to edit the machine's existing programs which had been written 'upstairs' by engineers and programmers. In many cases, this skill had been acquired through trial and error, spurred on by the natural curiosity which was common to virtually all the skilled men who were interviewed.

In the printing industry, as in engineering, craft-based skills are male-dominated, rooted in a lengthy apprenticeship, and reinforced by strict union demarcation. Although the nature of the work in the two industries is different, there are some similarities about the new technology which has been implemented in recent years. In both cases the operator's attention is directed primarily towards a control panel or VDU rather than to the machine itself, and the pace of work is more system-determined than controlled by the operator. Although new technology was evident in binding, guillotining and camera operating, the area of printing in which microprocessors have made the biggest impact is in typesetting and compositing. As in the case of engineering, the general perception was that while skills changed as a result of new technology, there was no sense of any overall loss of skills, the lessened craft input being more than compensated for by the acquisition of new skills which enhanced their owner's marketability. Printers were quick to point out that the new technology meant for them an increasing reliance on intellectual rather than physical skills.

The third category of effects of new technology occurs where the worker perceived that he has gained something but also lost something. One example comes from shirt-making, where the installation of automatic machines had abolished all of the traditionally skilled work required of the seamstresses, and had replaced it with the requirement to load and monitor the machines. This was accompanied by a fall in the quality of the final product. The sense of loss experienced by the seamstresses may in some cases have been compensated by the liberation from what was, after all, a physically demanding task. In the electronics industry, too, there were some workers who clearly missed at least some elements of the old electro-mechanical technology. These workers were generally located in the testing areas, where not only had much of the fault-finding been taken over by electronic devices, but the skill required to rectify the fault, once identified, had been replaced by simply discarding the faulty component. In this case, the sense of keeping abreast of the state of the art in the fast-changing industry was balanced by a sense of the job having been deskilled and made less satisfying.

The electronics industry also provides an example of cases where workers felt that the advent of new technology had made their jobs less skilled and less rewarding, without any other compensating improvements. Long-serving employees had seen the nature of their jobs change from being part of the manufacturing process to the straightforward assembly of electronic components. Such a feeling was exacerbated by the adoption of automated PCB populators which removed the need for teams of hand populators on the assembly line. In one company, workers were queuing up to be transferred from the assembly-line to testing, or to virtually any other job, simply because the assembly-line work had become so boring and monotonous.

It was in the electronics industry, too, that the process of skill-polarisation as the result of new technology was most clearly evident. This is the process in which the gap between the most and least skilled members of the workforce widens. In the previous generation of technology, those who tested and attempted to rectify the numerous faults in electro-mechanical machines could reasonably make suggestions for improvements to engineers whose level of product knowledge may not have been much greater than their own. Today, the same workers now supervise machines which test or assemble electronic components, leaving little opportunity for them to exercise their skill or understanding. At the same time, the technical complexity of the final product increases the skill required of the engineer or the technician so that the gap in skills and communication between the two groups of workers continues to grow.

Finally, in many industries there is a class of worker whose function is threatened not just with deskilling but with total elimination by the advent of new technology. This is the supervisor, variously known as foreman, leading hand, charge-hand and any one of a host of other titles according to which industry he or she is in. They have in common the attribute that they are the link between management and workforce. Twenty years ago, a supervisor was a person who, through experience, had gained more technical competence than other staff, and who thus qualified for promotion. But the new technology can be learned very quickly. The other major function of the supervisor, especially in manufacturing industry, is as an organiser of the production process. But computerised production control systems are rapidly eliminating this function too. In the light of these developments, the typical supervisor has increasingly become devoted to personnel and motivational functions, and some supervisors are now being asked to acquire formal supervisory skills for the first time. This is quite the reverse of trends expected by earlier writers towards a more technically-oriented supervisory function (Rothwell, 1984). If, however, this personnel and motivational function is seen as a normal one for middle or higher management, then it seems possible that new technology may eliminate the supervisory function altogether.

CONCLUSIONS ON SKILLS

Skills are multidimensional, and the evaluation of each element is highly subjective. New technology, while making some skills and some crafts redundant, creates new functions and thus new skills and new occupations. One cannot conclude that there is a long-term trend towards less skilled occupations unless one defines 'skill' as 'individual dexterity'. But dexterity is not the only dimension of skill, and an argument can be made that there is a tendency for the overall level of skills to increase with technical progress. In particular, it may be argued that there is a tendency towards the expansion of public services employment, and towards the profession-alisation of public service jobs.

OTHER HUMAN EFFECTS OF NEW TECHNOLOGY

The overwhelming impression which we received from our survey was that factory workers and office staff alike were favourably disposed towards new technology in advance of its deployment, and that exposure to it usually had the effect of reinforcing this favourable attitude. The main evidence supporting this impression was the queue of volunteers in almost every workplace anxious to be the first to be retrained on the new machines.

There were two main pre-deployment fears: a personal inability to cope, which almost always disappeared with experience; and secondly, the fear of losing one's job. This too was weakened *ex post* but generally persisted. There was also a definite feeling of inevitability: that new technology represented 'progress'; almost a manifestation of the heroic materialism of the Victorian age.

One post-deployment attitude which is worth noting was a sense of frustration when the new system failed to work. The psychological problems of frustration and helplessness when a system failure occurs are particularly prevalent when an entire group or section of the workforce is dependent on a central computer system. Such a situation can arise in the financial sector. When the central system malfunctions, the staff affected are powerless to rectify the fault, and if the previous system of manually-maintained paper records has been abandoned, there is literally nothing they can do other than fend off irate customers until the system is functioning normally again. The sense of frustration felt by staff in these circumstances has been christened 'the dependency syndrome'.

Our survey also left us with the impression that resistance to change on the part of the workforce was greatly overestimated by management. Many anticipated problems simply failed to materialise. Indeed, the widely held belief that 'no one likes change' must be seriously qualified. It seems true that people do not like *uncertainty*, but if they know that they are not only to continue in employment, but also that they will be retrained so that their personal marketability improves, then change becomes much more attractive. Once deployment has taken place, people adjust to new technology very rapidly. Although younger workers do seem to have a more positive attitude to new technology in advance of deployment, older workers do adjust if given the proper training and encouragement. While age was a factor widely believed by young people to influence acceptability, we came across numerous examples which contradicted this proposition. There was similarly conflicting evidence about the belief that the level of education influences the likelihood of an individual accepting change in the form of new technology. The proposition that a higher level of education renders an individual more likely, other things being equal, to adjust to change, seems to be more strongly held by those who are relatively less well educated.

A final factor influencing favourably the acceptance of new technology is the improvement in working conditions which is associated with it. While this improvement is more marked in some industries than in others, we never heard of any instance in which the physical working conditions had deteriorated as a result of the deployment of new technology. In the manufacturing industries, new technology means that work is less physically demanding than before, as well as being cleaner, quieter and safer. For example, although doing essentially the same job as a traditional lathe, a CNC machine is able to do so without the same element of grime for the operator: and many machines have an automatic swarf removal system. Improvements in working conditions of this sort are keenly felt by the operator; precisely the same effect is encountered in printing with the move away from the 'hot metal' process.

While fears of job losses and the inability to cope with new technology were the most frequent fears of the workforce *before* deployment, these feelings tended to give way, after deployment, to two quite different ones: an increased sense of pride or esteem in operating sophisticated equipment, and a feeling that new technology represented an investment in the individual worker's future employment security. The first of these feelings has two strands: a sense of pride in being part of a company which is recognised by others to be in the forefront of progress, and the self-esteem which can arise from being fully conversant with some aspect of new technology. There was no doubting the feeling of pride experienced by many people who had undergone training on computerised equipment, and this frequently led to a feeling of enhanced status, even where this was not formally confirmed by promotion or a pay increase. This feeling of esteem could also be shared by management, even when they had no personal contact with the new technology.

Workers often harboured short-term concerns about the effects of new technology on employment, while recognising that, in the long run, technology helped to secure jobs within a company as a result of its ability to improve that company's competitiveness. Frequently, the individual operator held optimistic views about his own job prospects and the long-run benefits of new technology for his company, while at the same time feeling pessimistic about the industry's outlook as a whole, and about the overall effect of new technology on job levels and skills in general.

For a very long time, technical progress in all its manifestations, but particularly in the field of production, has been held by critics to be 'dehumanising'. This criticism has perhaps three elements. First, it is alleged that the individual will become the servant of the machine, and be forced to work at the machine's pace, as for example in Charlie Chaplin's 'Modern Times'. Eventually, it is alleged, the individual will become merely a cypher in the whole system of production; and ultimately, human relationships amongst the workforce themselves will be totally destroyed. In our survey, we found isolated examples which could be used to support each of these propositions, but there was no general support to be found for them.

To begin with the faster pace of work; it is undeniably true that in the overwhelming majority of cases new technology has the effect of increasing the rate of output, whether of goods or of services per unit of time, by many orders of magnitude. In only one or two cases, however, were we told that this resulted in a less relaxed style of working. For example, one worker said, 'The machine is always waiting on you'. But much more often, particularly in the financial services sector, we were told that the volume of work undertaken by the new machinery released the staff for more interesting activities such as talking to customers, or thinking about their work.

On the second point, we came across only one instance

where employees felt that they were treated with less consideration by management as a result of the deployment of new technology. This was in a finance company where a system of personal relationships which had been established by clerks formerly working for particular typists was broken up by the deployment of new technology, and all typists were thrust into a common typing-pool. This reorganisation produced considerable distress and hostility. However, it was not a necessary consequence of new technology, and the conclusion must be that whether or not there is any deterioration in the personal relationship between manager and workers depends entirely on the management, and not on the machinery.

The last criticism has been well expressed by Mishan (1969):

> the technical means designed to pursue further material ends may produce a civilisation uncongenial to the psychic needs of ordinary men. A civilisation offering increasing opportunities for rapid movement, titillation, research, effortless living and push-button entertainment does not compensate for a deepening sense of something lost: of the myths perhaps, on which man's self-esteem depends: of a sense of belonging; of the easy flow of sympathy and feeling between members of a group; of the enduring loyalty that comes only from hardships borne together. (p. 78)

It is the feelings expressed in the last two clauses of this quotation which are most appropriate to our present discussion. A number of workers in quite different environments, some in factories, some in offices, told us that they *did* miss the comradeship which had existed with the old technology, in some cases simply because now there were fewer people around. We were told that there was less laughter or joking, and in another case that computers were too quiet to permit whistling! Such views, however, were by no means general, and were often offered as an afterthought in response to persistent questioning about the negative effects of new technology. They seemed to be the characteristic reaction of particular individuals scattered across a wide range of circumstances, rather than the common view held by a number of people in one particular location. The only evidence of dissatisfaction about a particular circumstance which was held by more than one individual arose in the aforementioned case of the creation of a typing-pool. In all other cases, the response to new technology was overwhelmingly favourable. When asked, in conclusion, if they had any regrets about the passing of the old technology, the answer 'no regrets' was invariably delivered by respondents with heartfelt conviction.

PART III

Challenges

8 Challenges to Firms

In the global marketplace, where innovation is a significant element of competition, firms are challenged to adapt or perish. If they do not innovate, their rivals will steal an advantage on them, in terms of cost or product quality, an advantage which once lost may never be recovered. Paradoxically, therefore, a firm's desire for future security depends in part on its willingness to sacrifice some present security by taking the additional risks inherent in the decision to deploy new technology.

There are many firms, mostly but not always large firms, who, as a matter of course, produce their own new technology, usually in the form of new products; rarely do they develop their own new production processes. Such firms devote a significant fraction of their earnings to expenditure on research and development, and many of them are household names. The huge majority of firms, however, do not fall into this category: for perfectly good reasons, they are quite content to buy new technologies off-the-shelf, as it were, on the open market. It is with such firms that this chapter is concerned.

Whenever a new technology comes on the market, a few firms will rush to adopt it, almost for its own sake, regardless of other considerations. At the same time, there are a small number of firms who are content to survive on second-hand equipment, just as there are households who in their lifetime will never buy a new car. In between, the great majority of firms are faced with recurring decisions about what to buy and when to buy it. The decision to introduce new technology is more than just a routine investment decision, it is an act of innovation, and in the last decade studies of industrial innovation in all its aspects have expanded rapidly. In particular, the characteristics of successful innovation have been sought in terms of management characteristics, organisational forms, and the 'culture' of firms. For some, such as Drucker, successful innovation is a game with identifiable rules that can be learned by anyone. For others, it is up to governments to accelerate the growth rates of their economies by persuading firms to take up new technologies faster. For them there is the frustration of the old adage: 'You can lead a horse to water, but you can't make it drink.' Firms themselves are more likely to respond to market forces, to the pressures of competitiveness, rather than to government exhortation, even when that is sweetened with subsidies. For many firms, of course, the response to market pressures comes too late, and the same recession which finally convinces them that they must after all adopt new technology if they are to survive, wipes them out.

For still other commentators, there is some universal recipe for successful innovation which can be found by studying the recent performance of firms which have innovated successfully, by trying to identify their common characteristics. Our own view is that there is more than one recipe for success. Not only will this recipe differ according to the size of firm, the nature of the industry, the level of the technology, but even among firms which are similarly circumstanced, different strategies can bring survival and prosperity. In particular, we believe that, as circumstances are changing rapidly, the 'right' decision can never be known with certainty in advance, and that there is an inescapable element of luck in the outcome of any investment decision. Of course, we can identify some key factors which are always likely to be important, and we have mentioned some already: size of firm, nature of industry and level of technology.

There are other factors which do not on the whole seem to be very important in the process of innovation. National differences for example: the opportunities for innovation and the constraints upon it are often similar in different countries, but different in different industrial sectors. New technology may have sectoral, but seldom today has national frontiers.

Another factor which we found to be unimportant is access to market information. Whether through the trade press, international exhibitions, visits to other factories, travelling salesmen, or some combination of all of these, we found that even the most 'backward' firms were very well aware of the costs and performance characteristics of the technology which was open to them. Whatever their reasons for not adopting the new technologies as their competitors had done, it was not lack of knowledge of the possibilities.

This chapter is not about the challenges to firms which surround the decision to invest in new technology. It takes as given that such a decision has been made, and focuses on the question: what are the challenges faced by the firm trying to make the deployment of the new technology as successful as possible? The chapter is divided into three sections. The first deals with the communication of information about new technology to junior management and to the workforce as a whole. It deals both with *what* to communicate and *how* to communicate. The second section deals with the human relations aspects of introducing new technology successfully, while the last section deals with the challenges which firms face in adapting their training and apprenticeship schemes to the requirements of new technology.

COMMUNICATIONS

In the field of communications, size of firm is critical. In large firms, formal channels of communications are inescapable. They may take the form of departmental meetings, briefing groups, newsletters, noticeboards, memoranda or any combination of these. In medium-sized firms, formal and informal channels of communication exist side by side, but usually the 'grapevine' is faster, more accurate, and capable of conveying precisely the kind of information which senior management would prefer not to be conveyed. In very small firms, it is possible to dispense with formal means of communication, and to rely entirely on individual face-to-face contact.

Most large or medium-sized firms use their existing communications procedures to disseminate information about the prospective employment of new technology. If the nature and extent of the new technology involves a major upheaval in the firm's traditional working practices then the existing procedures may be adapted to include special *ad hoc* meetings to deal with this particular subject. In the extreme case, a firm which attaches particular importance to the deployment of new technology, or to the associated organisational changes, may wish to underline this point by setting up a new set of communications procedures to run alongside the existing system. Indeed, a major change of this kind provides an opportunity for a firm to discover just how effectively its existing channels of communication are working.

WHAT IS TO BE COMMUNICATED

Information about the prospective employment of new technology in an office, plant or firm, should as far as possible satisfy five principles:

(1) The information should familiarise the workforce with the concept and workings of the equipment: the location, effects and timing of deployment should all be specified. The communication of such information is a fear-allaying exercise, whose purpose is to dispel uncertainty and anxiety. Perhaps because we all crave certainty, employees can never have enough information, but it must be specific. To give only general information may be worse than useless, since it not only will not dispel anxiety, but it may raise expectations which are later unfulfilled. Management should be careful never to oversell the benefits of new technology, as this can lead to workforce cynicism and dissatisfaction when it fails to deliver. Likewise, slippage in the deployment timetable must be avoided if possible. Workers and management are often motivated and excited by the prospects of new technology. If it fails to appear on schedule, they may very rapidly become demotivated and forget the initial training which they had received. Careful attention should therefore be given to the *quality* of the information communicated.

One finance company which we interviewed only got the introduction of its new technology right at the second attempt.

The first time the management made very confident claims in advance about the benefits of completely computerised systems of customer accounts. They did not, however, give the detailed attention to the implementation procedure and retraining which they later realised that they ought to have done. Thus, when the software proved not to be up to the tasks it was supposed to perform, the whole system collapsed. The experience left a legacy of distrust and hostility towards new technology amongst both staff and management, which took a three-year programme of rehabilitation to mitigate. Even when the second system was successfully working, the feeling of distrust of new technology within the company was almost tangible.

(2) The most important fear in the mind of the individual worker concerns his or her job. Early and specific information on the way in which the new technology is about to affect each individual is therefore required. Evidently a firm which is able to offer guarantees of no job losses, or at least no compulsory redundancies, is likely to gain more rapid acceptance for the introduction of new technology. A company which is large enough, flexible enough and far-seeing enough to be able to operate a 'full-employment' policy will have few fears to dispel.

(3) Staff will have fears about their ability to cope with the new technology. Therefore, information must be provided about the company's plans for the provision of training or retraining.

(4) Wherever possible, individual workers should be *consulted* rather than *informed* about the implications of the new technology for their particular task. This may appear like a counsel of perfection, and obviously the possibilities in a large firm are limited. However, in smaller firms we came across cases where workers' opinions were invited about the merits of alternative equipment, and in some cases prospective operators were allowed to specify a particular piece of equipment, subject to budget and other constraints. Even in a larger company, it is possible to take into account the preferences of the staff. In one finance company, although the clerks had no influence over the choice of VDUs, they were allowed to discuss the layout of their new offices, and to

choose much of the furniture. Thus, consultation is possible in large companies, especially if it can be broken down into smaller groups. Consultation to such an extent may not always be practicable, but the returns in tems of employee morale and commitment to the new technology are immeasurable.

(5) The importance of *feedback* should not be underestimated. Even where no prior consultation is practicable, employee morale will be much improved where the possibility exists for reactions to be passed up the line after deployment has taken place. Indeed, the post-deployment information process should be two-way. A continuous flow of information in both directions about new developments and possibilities can not only act to routinise the whole idea of new technology deployment, but it can enhance an employee's sense of responsibility. In this way, good communications can help new technology to be a catalyst for change, in the organisational and motivational sense, as much as it is the culmination of change in a technological sense. Of course, simply establishing the right channels for feedback is of little use if management interest is lacking. One of the banks which we interviewed set up a system of briefing groups, following their deployment of new technology, in order to improve communications. At first, the groups did no good at all, because management was not committed to the idea. They had to be trained to elicit feedback from the staff.

HOW TO COMMUNICATE

The need for formal channels of communication is quite different between large and small firms. The following remarks are therefore primarily addressed to large firms: problems with formal communications in small firms can usually be circumvented by informal methods, given an open attitude on the part of management.

The first point to be made is that the more *direct* is communication between management and workforce, the better. Whenever information is disseminated from the top down through intermediaries such as foremen, supervisors, shop stewards or other union representatives, councils or committees, invariably something is lost. Both the quality and the quantity of information are diminished when they pass through these traditional filtering devices. Of course, this point is not confined to communications about new technology, but applies with equal force to communications in general.

Fig 8.1 is a representation of the communication process which is particularly common in manufacturing. The strength and thickness of the flows illustrate the filtering role of the intermediaries. Even when a special working party is set up to tackle a particular exercise, there can be a loss of information, as Fig 8.2 illustrates. In our experience, it is preferable for management to bypass these channels and speak directly to the workforce. The preferred or direct communication method is illustrated in Fig 8.3. Too often, in the traditional communication process, when management believe that they have passed on adequate and accurate information, little of this information gets through, so that there are needless worker anxieties about their job and their ability to cope with new technology.

Banking, insurance, and other finance companies typically consist of a head office and an extensive branch network. Such an organisational structure makes the communication of information particularly difficult, and this was borne out by the interviews which we conducted. In most instances, the traditional communications method of memos and circulars was used to inform staff of impending new technology developments. In a few cases, an existing briefing group or meeting structure enhanced this procedure. In most of the finance firms in our case study there was virtually no face-to-face contact between senior management and staff at all. Just as noticeboards are an inferior substitute for direct contacts in manufacturing plants, so memos are an equally poor substitute in offices.

The general dissatisfaction of staff, outside head offices, with deployment procedures for new technology supported this view. However, there has in recent years been a growing recognition of the 'communications problem' and one of the leading British banks has taken some remarkable steps to try to redress the balance. Amongst other moves in a concerted programme of action, it has spent £2 million on video

The communication process — Pattern I

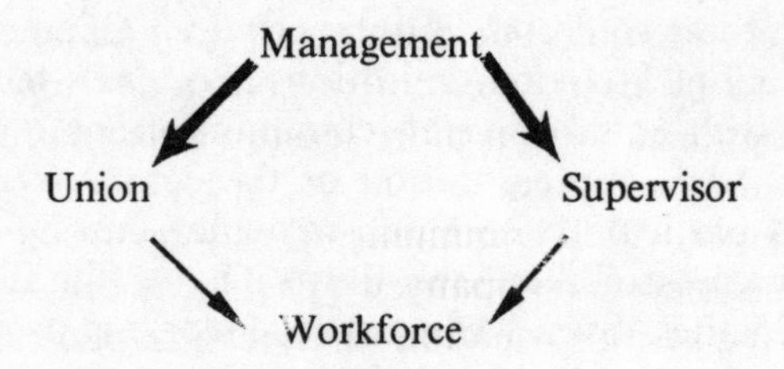

Figure 8.1:
Traditional communication process in manufacturing industry

The communication Process — Pattern II

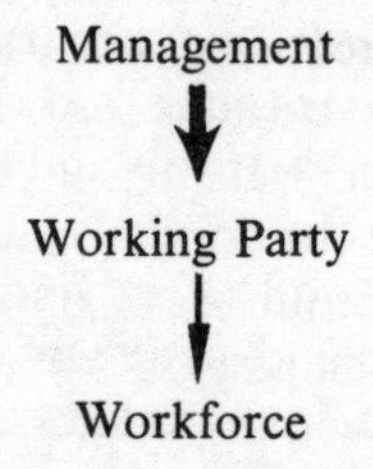

Figure 8.2:
Alternative communication process in firms unhappy with the traditional system

The communication process — Pattern III

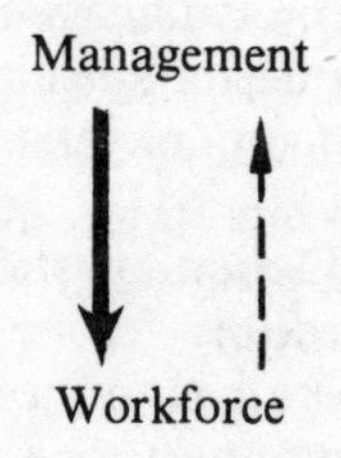

Figure 8.3:
The direct communication process. No excuses left for management not to produce the right information

equipment to be installed in each branch so that senior management in the head office can communicate directly with staff in the regions. The problem of communicating feedback has yet to be overcome, but of course developments in new technology make this a more feasible task than it might have been a few years ago.

The best effort in communications we came across occurred in an engineering company which by the mid-1970s had got into such a serious situation that its management realised that a major effort was needed to turn the company around if it were to survive. This meant organisational change as well as the introduction of new technology. Three years prior to the introduction of the new technology, the company launched a massive communications exercise. The unashamed purpose of this was to achieve a complete change in employee attitudes, not only to permit the required reorganisation and deployment of new technology, but also to make possible its commercial success. The workforce were formed into small groups and shown videos of where the company was going and why major new investment was required. This was followed by group discussions and feedback sessions at departmental level. Back-up newsletters were used to give the workforce answers to some of the questions they raised at the feedback sessions. When the new equipment was eventually deployed, operators were kept fully in touch with what was being done and why. All of these elements of the exercise were seen as being essential to achieve the continuing rehabilitation of the firm.

As a result of this communications exercise, industrial relations within the plant improved so much that a 30 per cent reduction in the workforce was achieved in 1977 without trade union sanctions. There was also evidence of success in cost-cutting. By the introduction of new technology and the reorganisation of working practices, the firm was able to improve its competitive position. In the world market for its principal product, it moved, over a three-year period, from being the twelfth lowest cost producer to the second lowest.

HUMAN RELATIONS

There is no special reason for identifying good human relations in a company, plant or office with the deployment of new technology. New technology is only one of the many elements in the successful performance of any firm which depends critically on good human relations. Our justification for introducing this very broad subject is that good human relations facilitate acceptance of new technology while, conversely, bad handling of the deployment of new technology can potentially sour an existing climate of good relations.

Human relations does not mean quite the same thing as 'industrial relations'. The latter phrase has come to imply an adversarial relationship between management on the one side and trade unions on the other. It was therefore a pleasant surprise to discover that a significant proportion of those interviewed in our survey did not understand what the phrase 'industrial relations' meant. The relationship between industrial relations and human relations can perhaps be illustrated in the following way. Suppose that, in any workplace, whether it be an office or factory, a secret ballot of the workforce results in a large majority not wishing to be represented by trade unions in their negotiations with management (while retaining the right to individual trade union membership). Such an outcome would preclude the possibility of any formal *industrial* relations; at the same time it might be seen as *prima facie* evidence for the existence of good *human* relations at that particular place of work.

Although such a test of good human relations can occasionally happen, trade unions are an important fact of life, especially in the manufacturing sectors, and increasingly in the public sector, of most of the industrialised countries of the world. And there is no doubt that they are there by the choice of the workforce. But within the last ten years the influence of the trade unions has declined. Membership has fallen, their advice has been ignored by governments, and in the workplace, managers have 'won back the right to manage'.

To many commentators, this decline in union power represents no more than a swing of the political pendulum, brought about by a combination of the recession of 1980–2 and a transient shift in popular opinion. Will the pendulum swing back again? In terms of the unions' influence at the workplace the answer is 'probably not'. As we shall see, there are long-run forces, including new technology, which are working to undermine the strength and influence of the unions in the industrialised countries. However, the actual influence of unions in each individual workplace will depend very much on the way in which individual managements handle human relations and, in particular, how they exploit the opportunities offered to them by new technology. This is one of the key challenges offered to management by new technology.

There are many people who believe that the human relations aspects of new technology can best be handled by an appropriately designed set of contractual relationships between management and unions. For example, Northcott, Fogarty and Trevor (1985), while agreeing that new technology deployment has hitherto been relatively troublefree, suggest that it may not always be so. They therefore recommend statutory standardisation of procedures, pointing to the nationwide agreements to be found in Germany and Scandinavia. British management has by contrast always fought a rearguard action against being 'tied down', as they see it, by contractual arrangements, preferring to see industrial relations arrangements developing 'organically', and dealing with issues pragmatically as they arise. 'Muddling through' can sometimes be a satisfactory way of proceeding, at other times not. It is also subject to the whims of intellectual fashion.[1]

It certainly would seem wiser to eschew, as far as it is possible, an extension of contractual relationships, since as Schumpeter pointed out long ago, any employment relationship which is based on contract only will not hold: some emotional element is necessary for it to endure. Getting not just the basic acquiescence of the worker to the deployment of new technology but also his or her constructive co-operation cannot be gained by a formula, however generous that formula might appear. The commitment of the individual worker can only be earned by management showing its respect for that worker. This principle may seem not only rather old-fashioned but may also appear so vague as to be devoid of substance; it may therefore be appropriate to illustrate its meaning with some more detail.

(1) Consultation as far as possible must be direct and personal

This can range from giving the operator a say in the choice of a new piece of equipment (which is possible in the case of a small firm) to senior management maintaining an 'open-door policy', a principle which can be, and is, practised in the largest firms.

(2) Breaking down social barriers

It is hardly necessary to point out how social class divisions destroy respect for the individual, since he or she is identified with the group to which he or she is assigned rather than as an individual person. We found many firms still operating canteens and other social facilities where 'blue-collar' and 'white-collar' workers were segregated. It became clear that these arrangements are resented although they have frequently become so taken-for-granted that one has to look closely to discover that they are in fact resented. New technology again performs an enabling function, this time breaking down social barriers in the workplace, since it is increasingly blurring the edges between traditional blue-collar and white-collar occupations, for instance, by making traditional manual tasks much cleaner.

(3) New technology makes possible increased machine monitoring of human performance

This is a highly sensitive development which is strongly resisted by workers and trade unions. In one non-unionised office we surveyed, the use of new technology for individual work performance measurement aroused such strong feelings that the climate of good human relationships which had previously prevailed was badly damaged.

(4) Other aspects

The extent to which a company is willing to encourage its employees to advance their careers, either by seeking formal educational attainments outside the firm, or by sending them on in-house training courses is very often an important indicator of that company's commitment to the individual. This statement has to be qualified by the recognition that, again, such things are more easily done by large companies than by small companies.

The evidence is less clear on the provision of recreational facilities and social functions. We interviewed one company whose recreational facilities could hardly be bettered. Indeed, it would be difficult to imagine a more comprehensive provision of welfare services to its employees than that company provided. While there were many workers whose commitment had been formed by this demonstration of the company's concern for the welfare of its employees, there were others who did not share such feelings and the climate of industrial relations in this plant was no better than in many other plants where such facilities were absent. Likewise, while personnel managers we interviewed were unanimous in their belief that such functions as company dances or golf outings, where management and workers had an opportunity of mixing together socially, made a positive contribution to breaking down social barriers and improving industrial relations, the reaction from the workers' side was mixed. In short, non-segregated social functions and activities can go a long way to making existing good human relations even better but are of little importance (or use) in breaking down barriers of suspicion or distrust when these are prevalent in the workplace on a day-to-day basis.

While respect for the individual may be an essential principle of successful human relations in the workplace, one cannot lose sight entirely of more mundane matters such as job security and wages. As far as job security is concerned, for the large company redeployment and retraining should be a possibility for at least part of its workforce whose existing jobs are eliminated by new technology. For the smaller firm, such possibilities are evidently limited, and therefore these functions must become the responsibility of government, especially if the rate of change of new technology accelerates. Except in the printing industry, no group of workers has appeared to expect their pay to be specifically and automatically linked to the deployment of particular pieces of new equipment.

TRAINING

Job security goes hand-in-hand with job flexibility; an employee can reasonably expect a degree of job security provided he or she respects the management's right to assign him or her to whatever job the firm requires at any time. This is the principle which lies behind the job security which many large Japanese firms are able to offer to their permanent workforce, amounting to about one-third of the total labour force in that economy. It is also the principle which lies behind the 'full-employment' policy of western companies such as IBM. In such companies, there is no room for job demarcation at employee level; as Northcott, Fogarty and Trevor (1985) point out: 'Japanese organisations do not usually have job descriptions ... boundaries between jobs functions or departments are left deliberately blurred, to encourage initiative.'

New technology, which blurs boundaries between technicians and craftsmen and between operators and maintenance men, directly encourages job flexibility, as well as making it essential if job security is to be maintained. To those few companies in the western world outside Japan which actively practise flexibility of job assignment, the development of new technology poses few problems. For the great majority of companies which are locked into demarcation, tight job descriptions and job boundaries, the advent of new technology sets a particular challenge.

This challenge is more of an opportunity for management than a threat, because it consists of two powerful incentives both working in favour of change. On the one hand, management evidently stands to benefit from increased job flexibility, and therefore has something to gain in bringing about the required changes. Likewise, workers can see that new technology is only a threat to their jobs so long as these jobs are viewed as unchanging. They, too, have a motive to accept change.

In the context of change, the attachment of any individual, whether he be manager, shopfloor worker or office employee, to one particular task can be seen to be defensive. It is also futile, since in a world of change any given task will sooner or later become redundant. In a world of rapid change, that time will come sooner rather than later. The inexorable effects of change can be seen very clearly in the apprenticeship system. For the past 150 years the traditional craft skills have been enshrined in an apprenticeship system. Many of these craft skills have been rendered redundant by new technology, and therefore the associated apprenticeship schemes are in disarray.

Everyone is agreed that what is required is to replace the long (up to five years), low-grade and single-skill training by a shorter, higher quality (more theoretical) and multi-skilled form of training. So far, few schemes have come forward to replace the old. This is partly a challenge for government, and partly a challenge for firms themselves. With the abandonment of the apprenticeship system, there is a danger that there will in future be a shortage of people with the necessary judgement to monitor the performance of the new computer-controlled equipment.

The training required goes beyond the absorption of purely technical information. One of the lessons to be learned from Japan is that 'anyone can display ability through self-development and mutual development backed by training and guidance'.[2] Over 90 per cent of shopfloor operators in car plants and electronics factories in Japan have continued their education until the age of 18. They therefore constitute a well-educated workforce, capable of listening to explanations of why new technology is necessary and of grasping its implications for the success of the company, and their future in it.

A related lesson from Japan is that companies should attach as much importance to their workforce as an asset for the future as they do to the equipment which embodies the new technology. 'Successful companies wherever they are located in the world all have a similar guiding ethos which, driven by the quest for significant world market share, treats all employees as the most valuable enterprise asset and therefore ensures they receive substantial in-house and external training on a continuing basis.'[3]

Training and retraining pose a particular problem for companies which do not have a mutual commitment with their workforce. When a company invests in a new piece of equipment, it can capture the benefits of that equipment for the length of its working life. However, when the same firm contemplates an investment in training a worker, that worker may leave the following week. Indeed, the training itself may well have increased his mobility by enhancing his market value. This is again a problem which is of varying significance for firms of different size. Generally speaking, larger firms can afford to invest more heavily in training than can smaller firms. Again, this is a matter to which governments must give attention since it is up to them to provide a mechanism for ensuring that what is desirable for the economy as a whole is also profitable for the individual company.

One of the remarkable opportunities which new technology provides to companies in the field of training is that, generally speaking, the ability to operate computer-controlled equipment can be learned very quickly—in a matter of days—with no previous skills or special aptitudes. This statement however has two important qualifications. First, although it may take only a few days for an operator to obtain a tolerable level of performance from a piece of computer-controlled equipment, it may take several months to attain optimal levels of performance. Secondly, it does require judgement based on perhaps years of experience to know whether a piece of equipment is performing properly or not. However, new technology itself can be used to make a little judgement go a long way: for example, in the textile industry a skilled weaver can now, with the help of computer-controlled monitoring equipment, supervise simultaneously the operation of eight or nine looms, where previously he might have been able to monitor only one or two by eye.

Perhaps because the operation of the new technology often appears to be simple, there is among many firms, a tendency for operators to be expected to learn by trial and error. Operators in a number of firms claimed that 'on-the-job' training was the best possible introduction to new technology. Certainly, in our experience, it did not necessarily result in an inadequate worker. However, firms which went through a more formal process of training, involving outside or internal lecture courses on, for example, electronics skills, ended up with operators who were able to investigate the potential of new technology much more thoroughly and systematically.

Whether hands-on training is given by an expert in-house or elsewhere, operators should have a chance to learn about new technology before they are expected to operate a machine in an expert fashion.

TRAINING STRATEGIES

The amount of resources which firms devote to the training of new recruits and to the retraining of existing employees appears to vary widely from one industrialised country to another. A recent survey by the Manpower Services Commission (1985) has estimated that United States companies spend up to 3½ per cent of their gross turnover on training, while Japanese companies spend up to 5 per cent. At the opposite end of the scale, British companies spend less than 0.2 per cent of their turnover on training. On average, just 14 hours' off-the-job training a year would be the norm in Britain whereas in Germany 35 hours' training off-the-job is considered good practice.

These figures relate to training generally, and not to training in new technology, but the association between the proportion of resources devoted to training and the relative performance of the economies concerned suggests something about the possible importance of training. This inference is supported by a comment, reported in Daly (1985), by a British supplier of CNC machines. In common with other suppliers, this company provides training courses for operators of new equipment purchased from them. The supplier remarked that although the courses were elementary in content, the average person attending was not very receptive to them, because of their low general level of training. As a result, said the supplier, he estimated that almost one half of the new CNC machines being sold in Britain were not used as efficiently as they might have been, because their full capacity was not understood.

These impressions were sustained by the evidence of our survey. Despite an investment in capital equipment which many firms saw as being essential for their survival in the market, the great majority had a complacent attitude towards the training of their operators. It appeared as if management believed that once the investment had been made their problems would be over. Few firms had what could be termed a training strategy, meaning by that a detailed and considered plan of how best to achieve an identified objective. Those that did tended to be the firms which had a clear view of the direction in which they wished to travel, and were intent on getting there, no matter how long it took them. They would be characterised as firms which involved all grades of employees in the training process, thus closing the gap between management expectations and employees' and machine capabilities, who set operators off on the right footing by introducing them to the machines in a working environment (often the manufacturer's premises) before they were introduced into the workplace, and who regarded the training process as ongoing rather than lasting for a morning, a week or whatever it took to work the machine 'adequately'. On-the-job training, however, is not to be dismissed. Many operators liked this type of training, and some firms did it well.

A training strategy does not necessarily take years to evolve or to complete. It does require direction on the part of management which filters through to the operator. Companies which wish to effect the sea change which new technology usually implies must be prepared to give their staff and workers time to become familiar with the new equipment and processes. To expect a miracle of perfection from the outset is to misunderstand both the technology and human nature.

NOTES

1. At the time of writing, it is fashionable in the management literature to speak of 'good communications', to examine the 'culture' of Japanese firms, and to try to compile 'tailor-made packages of motivation and control' (CBI/NEDO quoted in Northcott *et al.*, 1985, p. 37).
2. Handbook of the Japanese Union of Scientists and Engineers, quoted in Northcott *et al.*, (1985) p. 130.
3. Department of Trade and Industry (1985). This sentence recalls the famous one with which *Anna Karenina* opens: 'All happy families are everywhere the same: each unhappy family is unhappy in its own way.'

9 Challenges To Government

This chapter is divided into four sections. In the first section the possible responses of government to the effects of new technology upon society are discussed. The second section of the chapter deals with government policies towards the promotion and development of new technologies in industry. One of the most important tasks of government in the near future in all the industrialised countries will be to promote occupational mobility, and this is the subject of the third section. Closely related to occupational mobility is the question of training, another area in which government is widely recognised to have a responsibility. The last section discusses the challenges which new technology poses for the administration of government.

GOVERNMENT AND SOCIETY

In Chapter 1, we identified some of the foreseeable consequences which the introduction of new microprocessor technology is likely to have for society as a whole. To the extent that government wishes to have an influence over some of these consequences, then this poses questions for government policy. Answers to these questions, of course, depend very much on the view which the commentator takes of the proper role of government in society. In principle, a very wide range of views is possible, from a totalitarian standpoint in which all responsibility for the future direction of social life is vested in government to the opposite extreme of an ultra-libertarian viewpoint, in which the role of government in society is something to be minimised.

Let us forget ideology, and adopt instead the pragmatic viewpoint which is taken by Simon Nora in his famous essay, *The Computerisation of Society*. The desired relationship between government and society which is expressed therein is one which would probably be agreed by the great majority of informed opinion in the advanced countries. The functions of governments in this setting should be to try to achieve political stability, to facilitate change, and so far as possible, to limit the domination of the smaller groups in society by the larger and more powerful ones.

The early generations of computer technology have played into the hands of those whom Nora describes as the more powerful players in the social game, that is, large companies, whether privately or publicly owned, and elements of the bureaucracy. Traditional data processing was hierarchical and centralised. The new microprocessor technology, however, need not be an influence for centralisation: it can equally well be used for decentralisation. Whichever it is, is a matter of organisational choice. This is as true for society as a whole as it is for individual firms. but there are very powerful forces in contemporary society working in the direction of centralisation: 'the technical stratum's dream of rationality and the majority's desire for equality combine to expand the power of the state and its satellites', and Nora is in no doubt that government policy must throw its weight against centralisation in the interests of the smaller players in the social game. He emphasises that decentralisation cannot come about spontaneously: 'telematics can facilitate the coming of a new society—but it cannot construct it on its own initiative.'

Resistance to change is a common theme of commentators who lament what they see as the relative economic and political decline of contemporary European societies. Nora sees traditional institutions and attitudes as being among the more important obstacles to change, and he implies that it should be the function of government to help diminish or bypass these obstacles. Sclerotic is a word which has been used to describe the British economy, and Nora himself speaks of contemporary France as being 'une société bloquée'. Even if these were accurate assessments of the present state of affairs, governments in contemporary western democracies, unlike Peter the Great, Stalin or Hitler, cannot simply impose change on their citizens. What they can do is to allow new technology to do its subversive work for them. Nora again:

> The government cannot impose change; it has to create the conditions within which others will be able to produce it. By the institutional unrest to which it gives rise, data processing, properly used, may provide leverage for this evolution. (p. 119)

Amongst the means by which this leverage will be provided is increased information:

> by increasing transparency it (new technology) will raise the question of the security and privileges that issue from the shadowy zones of society. (p. 52)

Thus Nora envisages that new technology will eventually give rise to a much greater plurality in society.

Does new technology pose a threat to democracy? Let us consider two arguments which are commonly advanced. First, there is the familiar argument that governments will use new technology to collect, store and retrieve information about the private lives of individual citizens. This might include details of financial affairs, health, employment and educational records, which, if improperly used, could give a government, or persons in government, undue influence over an individual citizen. However, in the police states of the past and present the zeal and ruthlessness of the state security agencies does not appear to have been hampered by a lack of new technology. It is certainly true that new technology provides the *potential* for government abuse of power, which is why governments in most western democracies have passed Data Protection or Freedom of Information Acts to specifically limit the powers of the state. There is no threat to democracy so long as the citizenry do not surrender their political rights. If there is a future move in western democracies in the direction of the police state, it is more likely to come from the voluntary surrender of these rights by the citizenry in exchange, for example, for protection against terrorism than because of new technology.

The belief that advances in communications technology will inevitably lead to the increased power of the government at the expense of the individual citizens rests on the implicit assumption that the government is a single monolithic entity. In contemporary western democracies, this is not the case: governments tend to be composed of contending departments and agencies which pursue conflicting interests. So long as this state of affairs continues, the chances of totalitarianism are diminished.

Even if the democratic political process can be protected, there are many who fear that the low cost and high power of information technology will make possible a system of centralised economic planning, in which the production and distribution of goods will reflect the preferences of planners rather than the preferences of consumers. It is certainly true that new technology makes possible a vast increase in the quantity of economic information which can be centrally collected. The flow of information which has resulted from remote sensing in space gives an indication of what is possible. At the present time huge amounts of information are routinely collected and processed by individual firms and state organisations in the industrialised countries. Developments in communications technology make the cost of bringing together and storing all this information relatively low. The only remaining cost is of the human judgement involved in the coordinating decisions. There is a very real danger in the next two decades that the western world may be swamped by oceans of economic statistics. The problems of the official statistical agencies of western countries will be transformed from problems of gathering essential data within a limited budget to the problem of avoiding being overwhelmed by the sheer supply of economic information. In these circumstances, will central economic planning not become a practical proposition at last?

Even if the *quantity* of available economic information exceeds all expectation, it is still unlikely to be of adequate *quality* for the purposes of efficiently operating a command economy. This is because the decisions which are required to be taken, whether they be centralised or decentralised, in any modern economy cannot be based simply on past data, but upon subjective evaluations of the future, judgements which are likely to differ considerably from one individual to another.

For these reasons, we do not believe that new technology poses a threat to democracy—at least in western societies as they are presently constituted. It does however pose a number of other challenges to government, and these we shall now consider.

GOVERNMENT POLICY TOWARDS NEW TECHNOLOGY

Responding to political pressure for rising material living standards, most western governments since the war have attempted to influence both the rate of development and the rate of adoption of new technology in their countries. They have done this in the well-founded belief that the development of new technology makes a significant contribution to international competitiveness and thus to economic growth. There are a number of ways in which governments have tried to promote technical innovation. These include support for a patent system, the funding of technical education and of basic and applied research, subsidisation, regulation and public ownership of industries, and procurement, taxation and investment policies. Governments have given varying emphasis to different policies and while some policies—such as the educational ones—are vital, it is noticeable that those countries such as Britain and France where the government has taken a major direct role in the process of technical innovation, as measured by shares of R & D spending, have not been more successful in terms of economic growth than other countries such as Japan or West Germany where private industry has been the dominant force. If this is so, there may be a number of reasons.

In the first place, we may enquire into the motives for government intervention in the development of new technology. Granted that this is a field which is important for economic growth, what makes governments feel that their judgements about which areas of new technology development should receive the greatest investment should override those of others who, in this very complex area, might be thought to be better qualified to judge? Economists can give a number of technical answers to this question, justifying the overruling by government of market judgements, but an examination of the expenditure pattern of British and French governments since the war suggests that their motives are more political and more basic: a desire for national prestige and fear.

Just as some developing countries used to want to have their own steel mill and their own national airline, so Britain and France feel, apparently, that they must have their own aerospace industry, their own nuclear power industry, their own computer industry, and their own domestically-owned car industry. Yet Mathias (1983) reminds us that of the three richest nations of Europe in *per capita* terms—Sweden, Switzerland and Denmark—none has large-scale aircraft, basic electronics or nuclear power industries; two have no car industries, and the third only a hightly selective one.

The joint British-French venture in building a supersonic civil airliner, the Concorde, has proved to be a particularly costly flag-waving exercise. The British government has spent twice that amount in trying to preserve its last volume car assembly operation, and both French and British governments have spent heavily to keep alive their favoured domestic mainframe computer manufacturers.

It is not as if these were the only avenues of fundamental research in technology. There are many others which, if successful, offer the prospect of more useful results to society for a fraction of the cost. For example, desalination, a substitute for the internal combustion engine in motor cars and so on. But such ventures are not glamorous and do not contribute to national prestige or political independence.

While national prestige is a widely recognised motive for public expenditure, fear is less so. The argument was first put forward by Jewkes (1972), and runs as follows: with the spread of technology and expertise throughout the world, those countries which are already industrialised fear that their high standards of living are increasingly endangered by the competition of the poorer countries. America and Europe alike view with alarm the inroads made by Japan, and now other smaller East Asian countries, into their markets. What will be left, it is asked, for the now rich countries to make and to sell in order to maintain their present affluence?

Such fears are groundless, as economists know, but unfortunately politicians are less willing to listen to economists on this question than they are on questions of market failure. The result is that the industrialised countries are moving towards the protection of their most cherished industries. If anyone should think that this view of the factors influencing government policy towards new technology is either exaggerated or out-of-date, they need only read *The Computerisation of Society*. The author is a distinguished French civil servant of the highest rank, and yet it is difficult to escape the impression that he and his colleagues believe that the international economic system is a conspiracy to destroy the economy of France.

If prestige and fear are the basic motives which guide governments in their choice of investments amongst the different areas of new technology, it is scarcely surprising that these investments have seldom met with commercial success.

The fact is that governments are notoriously bad judges of market prospects. They tend to respond to political lobbying either by industrial or by regional interest groups. Indeed, as Rothwell points out, by far the greatest proportion of government subsidies for research and development in most countries has gone to very large firms. This may be equally true of procurement expenditure. It is precisely, of course, those larger firms that might be expected to have been able to support major projects themselves. It is therefore difficult to escape the impression that large firms get the lion's share of R & D subsidies from government simply because they have the resources, both financial and political, to devote to powerful political lobbying, resources which are denied to smaller firms.

The experience of new technology policies in the western countries in the post-war period therefore poses a challenge to government. First of all, it would appear desirable to formulate a new technology policy which has well-defined objectives. These objectives should be explicitly stated, so that for example every citizen could know if the government were determined to build up national production capacity in a handful of technology areas, such as microelectronics, biotechnology and information technology. Some citizens might then ask whether it was wise to adopt such objectives in view of the fact that every country at the present time seems to be concentrating on the same areas of new technology. Objectives having been defined, the next step is to formulate policy instruments which are equally well-defined, and which are likely to be appropriate. In this connection, it would seem imprudent to put all a nation's eggs in a few project baskets.

Since we have, in earlier chapters, commended the principles of flexibility and adaptability to industrial firms, it is only right that the same principle should be applied to policymaking. Policy initiatives can be introduced on an experimental basis. The experimental technology incentives programme of the United States Department of Commerce is an example of such a policy initiative. It allows for learning-by-doing, for modification, and if need be, for termination. Finally, governments should never allow themselves to become wholly and irrevocably committed to projects at an early stage of their development. Rather, they should develop each project a step at a time, appraising progress as they go.

OCCUPATIONAL MOBILITY AND TRAINING

The accelerating speed of technical progress makes it all the more important that economic systems should be adaptable, if the potential benefits of new technology are to be fully realised. In a competitive market system those economies which are slower to adapt will find themselves doubly penalised. They will not only lose the direct benefits of new technology, but they will also lose the indirect benefits, as their existing industries will be increasingly undermined by the competition of lower-cost, higher-quality products from their more adaptable rivals. The relative unemployment rates of each of the major western economies is a rough guide to their adaptability. The United States and Japan have much more adaptable economic systems than many of their Western European trading partners, notably Britain.

One of the major challenges facing all western governments,

as the pace of technical progress increases, is the capability to move workers smoothly out of old jobs and into new jobs. This task is not so easy as it sounds, since a significant fraction of the labour force in the older industrialised countries are ill-equipped to redirect themselves into new jobs. As Drucker has pointed out, many blue-collar workers in the advanced industrialised countries lack the skill, the self-confidence and the social competence to find another job unaided. What they do have, however, is the political capability to oppose or obstruct change. Unless society takes care of placing them in new jobs, they will tend to oppose anything new, including even those changes which will contribute towards providing more secure and better-paid jobs for themselves in the future.

What, then, can governments do? Their task is made slightly easier by the fact that the more vulnerable groups are readily indentifiable. Attention can be concentrated on those in the age group between 30 and 55. A potentially adequate early retirement provision can be, and often is being made, for those over 55, while those who are under 30 are capable of moving and placing themselves. It is the remaining one-third of the labour force to whom policy needs be directed. To the extent that the problems are to be found mainly in traditional industries, these are often concentrated in a small number of a large-scale plants located in a very few places. For example, three-quarters of all car workers in the United States live in only twenty counties.

Unemployment is a problem which arises only across organisational boundaries. One of the reasons for Japan's low rate of unemployment is that a significant fraction of its labour force is employed by organisations which regard them as a permanent or lifetime labour force, to be redeployed rather than made redundant. There are a small number of western companies which practise such 'full-employment' policies; they tend to be larger in size. For smaller firms, such redeployment may not be possible, especially in the short run, and here governments must accept primary responsibility for securing the smooth operation of change. Hitherto, many European governments have done little to facilitate the movement into new jobs.

It is not that there is any shortage of work to be done. In the non-market sector of advanced economies there is an unsatisfied demand for unskilled labour in the caring professions (health, social services), an effective demand which is limited only by the tax revenues at the disposal of governments, together with a certain amount of artificial barriers in the form of 'paper credentialism' which could be swept aside, given the appropriate political will.

In the market sector of the economy, the employer's demand for labour depends largely on the wage cost and on the skills of those seeking work. If those who were made redundant were to receive wage subsidies when they move into the new job rather than receiving payments of the same value for remaining unemployed, such an arrangement would surely be to the benefit of the worker, the employers and society alike. A similar scheme already exists for school-leavers in the United Kingdom. The extension of such a scheme to the labour force in general, if it were not to be prohibitively expensive, depends upon more and better information being available about the terms and conditions of potential new jobs on offer, and the circumstances of individual workers who have become, or are about to become, redundant. This is an area where new information technology can be helpful; by retrieving and coordinating employment records, it could make it much more difficult for employers or employees claiming a wage subsidy to defraud the system.

There are great difficulties in framing general laws about taxation and subsidies in advanced societies. The law must be the same in principle for everyone, yet it is a fact that everyone's circumstances differ in a number of ways. Any universal law for payment of subsidies or for the taxation of incomes must necessarily be arbitrary, turning away some who are 'deserving' cases, while at the same time allowing others to take 'unfair' advantage of the system. More finely detailed information can allow regulations to be more finely drawn, to the general advantage.

Changing jobs may or may not mean changing locations: it will almost certainly mean changing occupations, and thus skills. One of the great benefits of microprocessor-based technology is that it makes it comparatively easy for powerful skills to be acquired by semi-skilled and unskilled physically handicapped people. From now on, only a relatively brief training period will be necessary to allow any employee to use a small computer or even an 'intelligent' terminal.

The comparative speed with which the operation of computer-controlled equipment can be learned should facilitate a great expansion of training throughout industry. It is widely agreed that the most useful forms of training are within firms. However, there can be powerful organisational resistances to improvements in training practices within firms. These may include the inertia of some management: while one worker may quickly be trained on a new machine, it may require considerable effort to make sure that a significant proportion of the labour force is thus trained. There may be resistance from craftsmen who feel that they have no financial incentive to offer higher-skilled training to many of their workforce. Indeed, there may be a positive disincentive, since the enhanced skill also enhances the mobility of the trained worker. It must be a function of government to design the appropriate incentives to overcome each of these resistances to change.

However, governments have traditionally had a major role to play in industrial training, by providing educational facilities outside the workplace, such as technical colleges and other places of vocational education, where theoretical as well as applied skills could be learned. In the past, college courses have frequently been linked to the craft apprenticeship system, especially in engineering and other manufacturing industries.

Craft apprenticeships had three principal characteristics: first, each craft was defined by materials worked upon or by tools used. To these divisions corresponded an academic division of disciplines, e.g. electrical and mechanical engineering. Secondly, they were of long duration, normally up to four or five years, with up to as much as one year being devoted to full-time college attendance. Thirdly, they were based upon the notion of a once-and-for-all acquisition of lifetime skills.

New technology has completely undermined the basis for this traditional apprenticeship system. The essential skill of modern production is a process skill, e.g. a competence in both electronics and mechanical engineering. Secondly, specific skills, such as the techniques for operating particular types of new equipment, can now be learned very quickly so the justification for a long period of training in a particular skill has gone. This has been recognised in the printing industry in Scotland, with the recent replacement of the traditional full-time year in college by a thirteen-week block release scheme.

With new technology, what is required is education for adaptability rather than for specific performance. Adaptability requires not only a more broadly-defined approach to skills, but it also requires a higher general level of education. In the context of engineering, a higher general level of education implies at least three elements: (i) a better understanding of electronics and the ability to programme microprocessors, (ii) a higher level of diagnostic skills, and (iii) a higher level of understanding of production systems.

The importance of the general level of education of an engineering workforce was illustrated by the results of a recent study carried out by Daly (1985), who compared productivity levels between German and British engineering plants. The study carefully selected pairs of plants which were matched as closely as possible for the similarity of their machinery. It was concluded that the uniformly higher productivity of the German plants was attributable to the superior general educational level of the German workforce.

As we have seen, adaptability requires that training and work experience should be sufficiently flexible to make it easier for people periodically to take on new and different types of jobs. Even although, as we have argued earlier, the adoption of new technology should not diminish the total number of jobs, it will certainly change the kinds of jobs that are available, even within a single organisation. With change, there may be no scope for a particular individual to continue in the same job, but, given redeployment and retraining, there may be opportunities for new kinds of jobs. If the benefits of new technology are to be realised, i.e. if it is to be readily accepted in production, it is essential to tackle the mismatch between the existing skills of different types of people and the skill requirements for the different kinds of new jobs. It is the responsibility of the government to provide both the framework of incentives, and, where necessary, the educational facilities, to make the transition between old jobs and new jobs possible.

GOVERNMENT ADMINISTRATION

In Chapter 5 we alluded to a number of cases where attempts at the computerisation of government administration had proceeded less than smoothly. The problems of realising potential productivity gains in government administration by introducing new technology are made more difficult by the fact that there is no competitive market mechanism which is continuously stimulating the adoption of cost-cutting methods. There are of course, as in the market sector, problems with communications, training and industrial relations. And there is the final problem of the evaluation of the whole exercise, once completed.

In the case of the French government administration, Nora observes that if computerisation is allowed to continue to proceed in a piecemeal fashion there will be great inefficiency: notably redundancy, duplication, 'a surplus of resources here, unfilled requirements there', and budgetary constraints everywhere. Furthermore, should this procedure continue it is likely to lead to a future patten of government organisations which is unintended, and what is worse one which will not be able to be easily modified. What happens in government administration is especially significant in a country like France, where many other organisations, especially those in the public sector, model themselves on government forms of organisation. Clearly the situation poses a challenge to government policy-makers.

Two extreme solutions to this challenge are possible: one representing extreme centralisation and the other representing extreme decentralisation. While, as Nora noted, 'the technical stratum's dream of rationality and the majority's desire for equality combine to expand the power of the state and its satellites', any attempt to institute a wholly centralised information technology system would run into the problem that 'the will of the state is multiple, and most often incoherent, because it is expressed by the infinite number of basic administrative units'. On the side of decentralisation, there exists in France what Nora calls a 'political affirmation' or 'an anti-state attitude based on principle, the defence of certain private interests, and bucolic nostalgia'. It is evident from his choice of words that Nora does not take seriously any present proposals for substantial decentralisation of government administration. However desirable they might be from a political point of view, they are simply not practicable.

The response which Nora himself puts forward to the challenge which he perceives is to propose the establishment of a small and flexible unit, recruiting staff entirely on secondment from other departments to avoid the rigidity and empire-building which would otherwise arise. The function of this unit would be to 'analyse and forecast the evolution of the government's functions'. To prevent its fulfilling this task in a purely academic way, it would be tied into data-processing and the computerisation process. Composed of a few high-level personnel for its planning activities, the unit should not exert any coercion. Its task would be to analyse, warn, alert, propose and persuade. Its function would be to 'induce the public authorities to question themselves about the future of their departments'. The simplest method on intervention would consist in assigning a few data-processing specialists during the period of implementation of a large project.

Nora does not even touch upon the critical question of whether this solution—assuming it to be a practical one—would not lead to computerisation and reorganisation taking place according to the interests of the civil servants themselves rather than in ways which would serve the interests of the taxpayers or of the consumers of government services. Perhaps because Nora is a senior civil servant himself, he does not question the altruism of the civil service. Whatever the situation in France, observation of governments in other countries makes one a little more sceptical. Throughout the industrialised world, a number of serious challenges are posed to the political masters of civil servants, whether these civil servants work in departments of central government, regional governments, local authorities, or semi-state agencies.

These include the following: (1) Without the signals and pressures of competitive market forces to help them, can heads of government departments identify the most 'productive' or cost-effective or fruitful ways of introducing new technology into their administration, as opposed to the superficially attractive ones? (2) Assuming that this problem can be overcome, can the same heads of department successfully overcome staff resistance to reorganisation, and can they provide the appropriate framework of human relations, retraining and communications which will facilitate the smooth and successful introduction of new technology? (3) If they achieve the first step, but are unable to achieve the second, in the face of overt or covert resistance on the part of elements of their staff, do they have the political will, and can they get the political backing, to enforce such changes, where the public interest clearly demands it?

Solutions to these challenges form the main tasks for future governments embroiled with new technology deployment. As yet little evidence exists as to their ability to handle such organisational upheaval, or to their willingness to face it head-on.

10 Challenges to Trade Unions

This chapter begins with an account of the declining influence of trade unions in the industrialised countries, as evidenced by their lack of success in realising their stated objectives so far as the deployment of new technology is concerned. New technology itself poses some particular problems for trade unions, which are discussed in the second section of the chapter. This is followed by two sections which, strictly speaking, do not deal with trade union matters; they concern two other groups in society whose function seems to be threatened by the arrival of new technology, namely supervisory staff and professional economists.

THE DECLINING INFLUENCE OF TRADE UNIONS

The declining influence of trade unions in the industrialised countries since 1979 has nowhere been more clearly seen than in their failure to have any significant influence on the process of new technology deployment, despite the enormous leverage which the potential obstruction of deployment has offered them. The extent of the shortfall between the trade unions' ambitions and their actual achievements in this direction can be measured by examining the attitude towards new technology formally adopted by the British TUC in 1979, and embodied in subsequent documents by the TUC and individual unions. This policy can be summed up in the following points:

(1) Unions should be involved from the start, 'at the

planning stage, with common information available to both management and unions'.
(2) No change should be made except by agreement: there should be a *status quo* clause.
(3) There should be formal new technology agreements.
(4) No redundancy.
(5) Shorter working hours.

Even the CBI, representing the most conciliatory elements of British management, at a time when that management was widely characterised as being cowardly and incompetent, refused to accept three of these key points. It did not accept the principle of the status quo clause that change should only be by agreement, nor did it accept the principle of the 'no redundancy' clause, and it rejected any necessary specific link between the shortening of hours and the introduction of new technology. Very few new technology agreements have in fact been signed, even with 'friendly' employers such as Labour-controlled local authorities. Of those which have, many are empty of specifics, and others have since been abandoned or effectively ignored by rank-and-file members.

As far as individual employers are concerned, the typical experience has in fact been that unions have simply been ignored. Like the banking unions they have stood on the sidelines, wringing their hands: 'new technology, its implementation and development in banking and finance, continues to progress in the United Kingdom at a very fast rate without negotiation and without consultation with BIFU' (BIFU, 1982). The same pattern is true in Germany.

The reasons for the weakness of the unions in the matter of new technology are not hard to seek. As we have mentioned before, workers, including workplace union representatives such as shop stewards, actually like new technology, and are reluctant to do anything to obstruct its deployment. They see the role of their union on matters of new technology as being, in general, restricted to such issues as health and safety. Since the deployment of new technology is not a one-off event, but something which is likely to continue for many years to come, then, unless there is a sea change in workers' attitudes towards new technology, it is likely to act continuously to undermine the influence of the unions in the long run. There are two other factors which appear to be undermining the influence of trade unions in the workplace. One of these is related to new technology, the other is not.

In the United Kingdom, unions representing skilled workers, traditionally the most powerful, have hitherto been organised on a single-craft basis. Yet, the effect of new technology, especially in manufacturing, is to render redundant these traditional crafts and, instead, to create a demand for a multi-skilled worker. In other words, the occupational basis for much of traditional trade union organisation is being undermined. This process has only just begun, and can be expected to accelerate in the next decade.

The second factor working to undermine the influence of trade unions in the long run is that at workplace level amongst trade union representatives there is a substantial interest in what might be called a 'developmental' rather than an 'adversarial' approach to industrial relations. There are some signs that management may be winning the battle of loyalties between themselves and the unions for the hearts and minds of their workers, including local workplace union representatives. Some academic studies (e.g. Fogarty, Daniel and Brooks, 1985) bear out what recent events have shown, namely that trade union leadership at the national level has greatly overestimated the adversarial inclinations of their members. For example, in the continuing study of shop stewards carried out over a number of years by the Trade Union Research Unit at Ruskin College, Oxford, the great majority are found to show a general absence of hostility to private enterprise, to be motivated by a strong personal sense of responsibility, to be interested in management efficiency and the quality of working life, and to be willing to co-operate with management in solving problems of productive efficiency as well as industrial relations. We ourselves came across an instance where shop stewards had suggested to management that the deployment of new technology was overdue.

THE CHALLENGE OF NEW TECHNOLOGY

For trade unions, the challenge of new technology cannot be divorced from the host of other economic and social influences which are pressing upon them and leading them gradually to alter the pattern of their development. But in the near future technology seems likely to become an increasingly important member of the set of such influences. We can identify at least five separate challenges which new technology poses to the unions.

First, the job-displacement effect of new technology is bringing about a decline in the membership of individual trade unions; in many cases, this decline is projected to accelerate. Within such unions, servicing the membership on a shrinking financial basis has tended to alter the distribution of power within the union. Although, in some countries, unions remain unduly fragmented in terms of size, the advent of new technology has accelerated the tendency towards the amalgamation of smaller unions and towards the strengthening of central union bodies and international union co-operation.

Secondly, the deployment of new technology has led to an increase in the level of sophistication of trade union bargaining techniques. In some countries, the negotiating of technology agreements through supplementary collective bargaining procedures has become something of a specialised art, producing in Norway the phenomenon of the 'technology shop steward'. In Britain, the white-collar trade union APEX worked out as early as 1979 a negotiating strategy for its local officials to come to terms with the introduction of word-processing and other forms of office automation. The strategy recommended is set out in diagrammatic form in Figure 10.1.

A third problem posed for the trade unions by new technology is that it creates conflicts of interests within a given union. There may be a simple conflict between the employed members trying to preserve their living standards, and the unemployed members (or those who feel they may be about to become unemployed) seeking to preserve their jobs. Some commentators see a further split within the constituency of the employed. It is alleged that there are those who flourish as the 'aristocrats' of the new technology, and

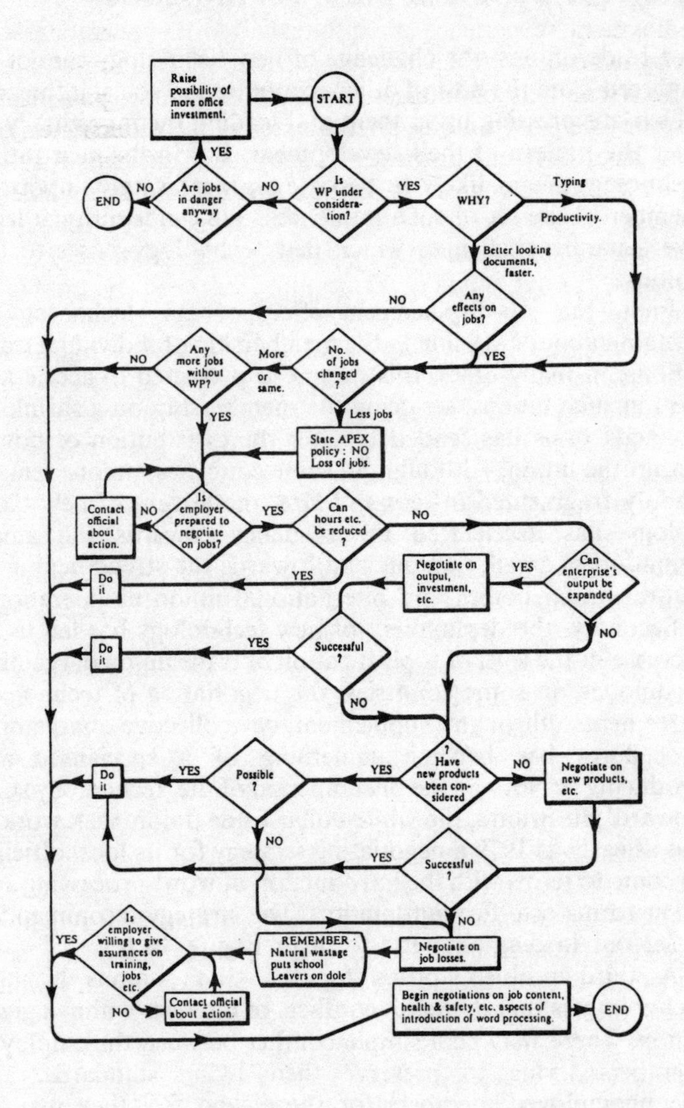

Figure 10.1
Strategy for the introduction of word-processing (WP) and other office automation
(From APEX, *Office Technology — The Trade Union Response*)

another group who, it is claimed, as a result of new technology deployment, experience more intense work, uncompensated for by increases in real earnings or reductions in hours of work.

Such conflicts of interest can also, of course, give rise to disputes *between* unions, to the extent that different unions may have different priorities. It may be significant that the Trade Union Congress in Britain has seldom been able to achieve its declared objective of having new technology agreements in multi-union plants. New technology also inreases the possibility of disputes between unions (and sometimes within unions), where unions are organised on a traditional craft basis. New technology erodes the traditional barriers between the work done by people of recognisably distinct crafts and occupations. We have already referred to the disputes in Britain between the blue-collar (AEU) and white-collar (TASS) engineering unions about who should program CNC equipment. Again in Britain, the trade union representing electricians (EETPU) has enthusiastically embraced the new electronically-controlled printing machinery, thus bringing them into spectacular conflict with the traditional print unions, (SOGAT and NGA). As the pace of new technology introduction quickens and the range of new technology expands, the opportunities for such demarcation disputes seem likely to grow. Such disputes are likely to lessen the influence of unions in society as a whole, if not in the individual workplace.

Lastly, while retaining their conventional defensive functions, that is defending the interests of their members in the workplace, the unions have to be willing and able to change their own organisation and methods in the light of changing circumstances. That this is more than just a self-evident truth can be judged from the tone of urgency which forms the following extract from a British TUC document:

> While changes that have taken place in the past in union organisation and union methods have been substantial and far-reaching, they have occurred as fragmented and uncoordinated responses to gradually changing circumstances. The pace of current changes will not afford the luxury of gradualism. Unions must, as far as possible, plan change. The task for each union is to build a sensible, flexible, strategy to enable it to adapt to the changing economic and industrial environment.[1]

How can unions adapt to change, as the TUC urges? Those unions which persistently set their faces against progress and change are certainly doomed to see their membership dwindle. One response is to go aggressively for 'sweetheart' deals with large companies, in which the union achieves single-union representation and in exchange gives guarantees against strikes and other forms of industrial action. If the union in these circumstances can persuade the company—typically a large multinational company—to accept a continuing responsibility for providing jobs for its members (even when their original job functions become redundant), then such a policy might well have a future. Significantly, these types of arrangement are more common amongst high technology firms. Not all firms in Europe or the United States may be willing or able to give such undertakings.

In the last chapter, we pointed out that, in the great majority of cases, it must be the responsibility of government in the industrialised countries to ensure the transfer of individual workers from jobs which are no longer necessary to ones that are. But for those who are as sceptical of government assurances as they are of the promises made by commercial companies, and for those who like to wear a belt and braces, some additional form of safety-net may be desired. Bearing in mind that workers do not on the whole seem to wish their trade unions to negotiate on their behalf concerning new technology, but do want unions to protect them in the event of potential management unfairness, it may be that there will be an opportunity in the rapidly changing society of the future for unions to act as organisations providing insurance to their members against the increasing risk of redundancy. It will be recalled that trade unions were originally formed as friendly societies to give their members the benefit of insurance against the risk of industrial accident and sickness.

If this were to happen, unions would no longer be attached to their traditional base of being defined by the tools their members worked with or the material they worked on. It would leave the way open for unions to compete for members across a wide range of industries and services. The more successful unions would be those who would offer the best services to their members. It would also seem likely that the more efficient the union the better protection it could afford its members against the possibility of management misbehaviour in a particular workplace. In this connection it should perhaps be mentioned that the better unions will not be necessarily those which have apparently the largest industrial 'muscle', in the sense of being able to inflict the greatest potential damage on the national economy. The evidence of recent disputes suggests that the ability to capture the support of that elusive phenomenon, public opinion, may be a more important element in winning industrial disputes in the advanced countries.

NEW TECHNOLOGY AND SUPERVISORY SKILLS

Foreman, leading hand, supervisor, charge-hand and any one of a host of other titles may be used to describe a member of a group whose precise function varies from industry to industry and from firm to firm. As we have seen, they share one attribute in common—they are the point of contact between management and workforce. The very ambivalence of this position can lead to problems in determining where the loyalties of this group lie (Rothwell, 1984), and forms the starting-point for an analysis of the two major challenges which new technology poses for the supervisor.

The first challenge is that of rapidly changing skill requirements, which comprises two elements; understanding new technology and adapting to it. The supervisor frequently plays a key role in passing information about new technology down to the workforce (with or without a corresponding role in the reverse direction), and is often instrumental in selecting and training the personnel who will operate the new machinery. In order to fulfil either role properly he or she must first of all be fully conversant with the new system and be informed of the reasons for its implementation and of its possible organisational and personnel implications. Interviews carried out with around 40 supervisors revealed that a substantial minority felt that they did not always possess the necessary skills to carry out these functions adequately, and that at least in part this was because of inadequate information from top and middle management. We now pose the question of why there may be a lack of information from the higher layers of management.

In smaller organisations it is relatively easy for there to be coordination between management and supervisor, with the latter frequently involved both in the decision-making and in the implementation of a new system. Problems may occur in larger organisations where decision-making is divorced from the implementation process. The decision to deploy new technology may be taken at a geographically remote head office, leaving management at the individual plant or office to make much of the preparation for deployment, and perhaps carry out the actual deployment itself. Where the local management know little about the new system they may feel their own position to be under threat, the therefore be unable or unwilling to pass what information they do have down to the level of supervisor. There may be a serious danger of appearing not to be fully in charge in the early stages, precisely the time at which the supervisor may have to cope with a stream of rumours from the shopfloor grapevine. A variation of this theme may occur when the impetus for new technology comes from a group of highly technically-oriented individuals who are the only ones who truly understand the proposed system. Where even the most senior management have limited real knowledge, the amount of information which trickles down to the supervisor can often be very little indeed.

Even when fully conversant with the new technology, the supervisor may be faced with the fact that his or her skill requirements will be radically altered by it. This change in the nature of required skills was perhaps most apparent in banking, insurance and finance. Twenty years ago the supervisor in a bank or building society was someone who by dint of long service had gained more technical competence than other staff, and who thus qualified for promotion. Today computerisation has made it much easier to grasp the technical details of the business, and although the range of financial services on offer has multiplied beyond recognition, even a relatively junior member of staff is expected to be knowledgeable about most of them. The role of the supervisor has now become more oriented to dealing with the personnel and motivational part of the job rather than its technical detail.

The second major challenge faced by the group is that not only may their skill requirements change, their role may disappear altogether. When asked why he was so enthusiastic about a new production control system, a senior foreman in a mechanical engineering company replied: 'Because it lets me get on with the real job instead of running around with bits of paper all day'. Clearly he saw his role as organising the production process and ensuring that the manpower requirements of his departments were met, a job which he rarely had time to fulfil because of the minutiae of his day-to-day activity. In the medium term he correctly perceived that the production control system would remove much of the endless paperwork from his desk, but when asked about expected future developments of his role as senior foreman, he said that he expected it to disappear in the form which he knew it. Eventually, he felt the system would be able to do much of the detailed planning which he carried out, and the need for a link between management and chargehands would be lost altogether.

Clearly this is only one example and it would be mistaken to believe that supervisors as a group will simply disappear because of new technology. But this does serve to illustrate the point that in many cases the basic task of monitoring and controlling the pace and flow of work can be carried out by the new technology itself. When supervisors do not become either technical specialists or human relations specialists they may find that the role which they have to play is dwindling rapidly.

CHALLENGES FOR ECONOMISTS

Despite more than 200 years of systematic development, the social science of economics in its present state is unable to provide a satisfactory framework for answering the type of questions which were raised in the first chapter of this book. While it might be unreasonable to expect it to provide conclusive answers to these questions, it is not unreasonable to expect it to provide a framework for attempted answers. However, contemporary economic models are inherently inadequate, even in addressing such apparently pertinent questions as: What is the effect of computerisation on unemployment, on foreign trade, and on the rate of economic growth of a national economy?

This deficiency of economics is not confined to questions relating to new technology. It is equally true of almost all pressing issues of economic policy concerning the advanced countries of the present day. For example, unemployment is observed to be at historically high rates in almost all the advanced countries of the world. Yet economics and economists in these countries have been able to offer very little in the way of explanations which are either convincing or even satisfying to laymen about either the origins or the cures of this phenomenon. Again, one would have supposed that economic analysis would be able to say something helpful to policy-makers about the consequences for growth of the national economy of a country joining the European Community. Yet whenever a country has tried to join the European Community, economics has found itself powerless to offer any conclusive advice as to the probable costs and benefits, and therefore adherence or, as in the case of Norway, non-adherence, has been very often a matter of blind faith. Why is economics so helpless to answer these kinds of question?

There are a number of features of economic analysis as it is presently practised which are clearly unsatisfactory, and some of these features are shared by other social sciences. Although it is evident to the man in the street that human behaviour cannot be compartmentalised, so that the behaviour of the individual human being is not determined on some occasions by purely economic factors, on others by purely psychological factors and on still others by purely social or political considerations, yet each of the social sciences insists upon isolating one dimension of human behaviour and analysing that. So, for example, when a social scientist is asked for his appraisal of a proposed housing project he will analyse the advantages and the disadvantages only of those aspects of the project which fall within the boundaries of his own particular discipline, as traditionally defined, and will ignore or underplay all the other aspects. Thus an economist will look only at the economic costs and benefits of the proposed project, while a sociologist will consider only the social aspects, and so on.

Secondly, despite Keynes' famous warning that 'a foolish consistency is the hobgoblin of small minds', contemporary economic theory has been built up upon a basis of satisfying one criterion and one criterion alone, and that is the criterion of consistency. No matter how unrealistic are the premisses from which it starts, or how absurd the conclusions with which it finishes, analysis in present-day economics is judged entirely by the consistency of the argument. Not surprisingly, it has lost touch with the changing facts of experience, so that it is quite common to find practitioners confusing the properties of their theories or 'models' with properties of the real world. They move from the world of the model to the world of experience, without troubling to see whether there is, in fact, any correspondence between a particular concept and its analogue in the world.

Not surprisingly, therefore, contemporary economic analysis is entirely exempt from a consideration of those specific factors of time and place which have a decisive influence on events. The given analysis or model is deemed to be capable of being transplanted from one situation to another without effective modification.

Finally, in a mistaken zeal for quantification, mimicking the natural sciences, contemporary analysis ignores the more subtle relationships in the life of a society that cannot be weighed and measured, but which may be more important to that nation's affairs than those things which can. Instead, economists in the last 50 years who have studied practical questions have devoted themselves to collecting and processing the available statistics, regardless of how appropriate these statistics are to the central issues of the question. This is a problem which is going to be aggravated by new technology. It is not just that, as we observed in the last chapter, there will shortly be a statistical explosion as data which are already locally located are transmitted worldwide, but new technology also facilitates the collection of new primary data.

Between March and September 1985, five survey teams interviewed members of 900 households in the Ivory Coast for a study of living standards. The teams used highly detailed questionnaires to interview each household in two or three sessions spread over two weeks. Normally statisticians would take years to publish as much information as the teams gathered, yet the group turned out a preliminary abstract of the results just two months after finishing the fieldwork. This was made possible by the fact that each of the survey teams was equipped with a personal computer. The computers not only helped to compile and correct the data: they were used to organise and manage the survey. Furthermore, the use of computers meant that the survey team required less highly educated people who required a shorter training than would otherwise have been the case.

While it is desirable that more primary data should become available at lower cost, there is a danger that this will encourage the collection of data which are more easily quantifiable.

For all these reasons, the contemporary economist who is invited to address himself to policy issues usually adopts one of two possible courses of action. Either he proceeds as if economic considerations were the only ones, and carries out an economic analysis with the characteristics just described, concluding by confidently making recommendations which are to be applied immediately to the real world on the basis of his highly artificial assumptions. Or, if he is exceptionally conscientious and is acutely aware of the limitations of economic analysis as it is presently constituted, he will feel unable to make any pronouncement on public policy issues whatsoever in his capacity as an economist.

The experience of either or both of these attitudes has led policy-makers and the general public to a considerable level of disenchantment with economists (and other social scientists), and to turn to 'practical' businessmen, consultants, civil servants, journalists and even politicians, for advice on the solution to these issues. As a result, to satisfy the demands of policy-makers, there has grown up a number of *ad hoc* evaluation procedures, such as Technology Assessment and Carrying Capacity which in some circumstances have become a substitute for economic analysis.

What, then, can be done? Just as the new technology is

making redundant the old craft skills traditionally based on working with particular tools or materials, so too is new technology making obsolete not-so-old but nevertheless conventional practices in economics and in other social sciences.

Since it is indisputable that the consequences of technical change extend well beyond the frontiers of any one of the social sciences (such as economics), and embrace all aspects of human activity, it follows that any method of approach which breaks down the barriers between the social sciences is likely to be a more fruitful one than the conventional compartmentalised approach. Under the weight of such phenomena as those set in motion by new technology, the old frontiers of the social sciences are changing, but the new frontiers have yet to be defined.

It is already evident, however, that what is required in place of the one-dimensional, formal but general and therefore empty approach of conventional economic analysis, is a method of analysis which consists of a broad well-balanced presentation of all the relevant facts seen critically from different points of view. If economics and the other social sciences should develop in this direction, then it is possible that they will become better-balanced, more effective and more practical than they are today.

NOTE

1. Employment and Technology, 1979.

11 Conclusions

In the preceding chapters we have reported on the economic impact of new technology, so far. The words 'so far' represent an important qualification, since the revolution in technology brought about by the microprocessor is still in its early stages. With this qualification in mind, however, it would be fair to say that the impact on the citizen as producer has been benign. The only malign effect hitherto has been on jobs, and that has been small and is capable of being corrected.

We have looked at the impact on citizens working in firms, in trade unions and in government. But what about the citizen as a consumer? Although this is not strictly within the terms of reference of our book, it would be unfortunate to conclude without some reference to this topic, especially as it has been the subject of gloomy prognoses by some eminent novelists.

Forster, Orwell and Huxley all anticipated that the same developments in technology which would make possible the abolition of material poverty in human societies would eventually destroy every vestige of humanity in the citizens of these societies. What is the substance of these fears? A number of strands may be identified. There is first of all the suggestion that human relationships in technically advanced societies will become drained of all emotion. This has been put very well by Mishan (1969) in the passage we quoted earlier (see p. 88).

Another charge which might be brought against advanced technology is that it may lead to the fragmentation of society, to a loss of social cohesion. Changes already taking place in the technology of public broadcasting (cable, satellite, etc.) mean that different people may get their information not from one or two sources, but from so many different sources that each one receives a different and incomplete fragment of the total information. More than 200 years ago Adam Smith observed that people in societies governed their ethical behaviour in a way which was calculated to enlist the sympathy of the impartial spectator.[1] But in a fragmented, compartmentalised society, there may be no spectators, impartial or otherwise.

Are we moving, then, towards a society centred on machines and away from a society centred on human beings? Are there any signs, so far, that such fears have any basis? There are a number of straws in the wind. The fact that any citizen attempting to correspond with a major organisation—electricity, or telephone or insurance, or credit card company—is treated with contempt if his circumstances and message do not fit into a form capable of computer processing. The fact that a major European oil company is now altering its operational procedures in retail distribution to conform to an existing computer program. The fact that production of VLSI chips is now becoming almost completely automated. The fact of progress towards fifth-generation computers which will have, amongst other powers, voice-recognition capabilities. Are we heading in the direction of the Orwellian nightmare, in which advanced computers will, as the head of the Japanese fifth-generation project has predicted, be able to 'eliminate distortions in (social) values'?

As Weizenbaum (1983) has pointed out, the answer to this question does not depend on the potential technical power of computing systems. It is a question of:

> the willingness of populations to surrender themselves to the "conveniences" offered by technical devices of all sorts, particularly by information-handling machines. Will they be eager to make the Faustian bargain by which they will have the things and entertainments they want in exchange for their right and responsibility to determine their own values?

To those who think this is a 'preposterous question', Weizenbaum goes on to remind his readers of 'the enthusiasm with which millions of parents turn their children over to the television set in order to escape having to bother with them', and of the fact that many people prefer to 'interact' with computers; some schoolchildren prefer them to teachers, and many patients to doctors.

The verdict must therefore be an open one: at this stage, the citizen as producer seems to have been able to cope quite well with new technology, but as consumer less satisfactorily.

Towards the end of 1981, the Japanese government[2] announced a ten-year programme to produce a so-called fifth-generation of computers. This objective would require progress in, and unification of, a number of heretofore separate areas of research: knowledge-based systems, human-oriented input-output functions, high-level programming languages, decentralised computing, and VLSI technology. It was claimed that if the goals of the research programme were reached, that machine intelligence would be sufficiently improved so as to approach that of a human being.

The Japanese announcement was immediately followed by an upsurge of activity in the United States and in much of Western Europe in those same fields of research. In Britain, as a result of government funding, much of the activity has been concentrated in the field of knowledge-based or expert systems (acceptable euphemisms for artificial intelligence). An expert system is a computer programme which captures the knowledge of the human expert or specialist in the form of rules, which then enable the computer to give an 'expert judgement'.

Despite these efforts, the fundamental theoretical and methodological issues have not been solved, and it seems quite possible that they never will be, since the effectiveness of such systems is based on the supposition that every aspect of nature, and most importantly of human existence, is precisely describable, and that all human knowledge is sayable in words.

However, even if the artificial intelligentsia's vision of thinking machines remains a mirage, and the ambitious efforts of the fifth-generation research programme fall short of their targets, it does seem likely that the research efforts expended in these areas of new technology will lead to the development of computers which are even faster, still more reliable and with a wider range of capabilities than those which exist at the present time.

The important consequences of the era of accelerated

technical progress which all the industrialised countries are now entering are not primarily technical but economic and social. The new technology has enabling effects, not determining effects: it creates opportunities which enable individuals to realise potential benefits. Sometimes these individuals will be acting on their own behalf, at other times on behalf of firms, trade unions, departments of government or other collective groups.

New technology also has subversive effects. It undermines existing technology, and therefore existing ways of doing things. Thereby it threatens the livelihood of large numbers of individuals, who may be quite unaware of that threat. To what extent the damage can be minimised is also a matter for individual and collective decision. Thus the effect of new technology is to pose challenges to individuals and to organisations in advanced industrialised economies. The nature of the challenge differs according to the context, but in each case the principle is the same: the decision-maker must realise as far as possible the potential benefits offered by new technology, and at the same time attempt to limit the disruption which it causes to himself or to his organisation.

Throughout this book we have used microprocessor technology to stand for all forms of new technology. The microprocessor is not the only technological innovation of recent years, nor is new technology the only cause of change. But new technology is perhaps the most important source of economic growth in the contemporary industrialised world, and it is likely to become even more important in the near future. And the microprocessor is certainly the most pervasive example of new technology in the industrialised countries at the present time.

If new technology is the most important source of economic growth in the western world today, then undoubtedly the dominant force which is propelling the industrialised economies along their growth paths is competition. Firms compete by innovation. They either adopt a new process of production which will allow them to produce the same goods as their rivals at a lower price, or else they try to adopt a new or better product which will improve their output and market share at the expense of their rivals.

The process of competition by innovation is a process of creative destruction. New machines and other forms of capital such as buildings and vehicles are created, and new jobs are created to go with them. At the same time, some forms of capital and some jobs associated with existing technologies are destroyed. Existing capital and existing jobs are not, of course, destroyed physically but only in economic terms. Their market value is severely reduced and in the extreme case it is reduced to zero.

Both the creative and the destructive effects of the deployment of new technology have direct as well as indirect components. The indirect effects may be more important than the direct effects, but they are much more difficult to identify. Surveys have shown that the direct job losses in factories and offices where new technology has been introduced have hitherto not been as great as had been expected. Although there is no truth in the old idea that technical progress must inevitably bring about a fall in employment in the world as a whole, it is quite possible that there may have been considerable indirect job losses, caused by competition and mismatching, in particular countries or regions, in particular industries or in particular occupations. And as the rate of progress increases, so, too, may the rate of job loss.

Competition by innovation leads to the under-utilisation of some specific forms of existing capital which are rendered redundant. Likewise, there is unemployment of occupationally specialised or regionally immobile workers. When these circumstances arise in a country at a particular period of time, this is sometimes mistakenly taken to be a sign of a temporary period of economic stagnation in that country's economy. In fact, this should be interpreted as evidence of a period of unusually rapid economic progress.

New technology can be unsettling in other ways. By abruptly lowering or abolishing barriers to entry to an industry, it can not only sweep away monopolies, but also undermine the position of dominant firms within that industry. The evolution of a particular technology, passing through various stages of its development, can at some times confer advantages on particular firms or particular locations or particular groups of workers, while at other times subsequent developments can remove these advantages. For example, the development of large-scale integration (LSI) technology in the 1970s made possible a mass market for chips and created a demand for semi-skilled labour, often in developing countries, to assemble and test them. The subsequent development of VLSI, and of custom-built chips, is shifting the demand to automated, capital intensive methods of production in advanced country locations. Inequalities between different groups of people and different locations can be reinforced. Growth can therefore be very disturbing, but this is inescapable. Without growing pains there can be no growth.

If firms do not adopt new technology when it becomes available to them, their rivals will sooner or later steal a march on them. Where there is competition, firms are therefore challenged by the advent of new technology to adapt or perish. Paradoxically, the key to the realisation of greater long-term security for a firm in an industrialised economy is its willingness to sacrifice some short-term security by taking the risk of investing in new technology.

If a firm is to deploy new technology successfully, good human relations are essential. These depend on job security, where that is possible, which in turn implies opportunities for retraining. Good human relations in the workplace also require that consultation should be direct and personal. Communications which are filtered through management, or union representatives or even a committee structure, are seldom satisfactory. Work people can never be given too much information, but it must be accurate: disappointed expectations can be fatal. So far as possible, there should be a feedback or consultative element in the communications process, even if it is only about the colour of the office furniture.

One of the unexpected findings of our survey was that workers, whether on the shopfloor or in the office, are usually favourably disposed to new technology in advance of its deployment. Actual experience of operating new technology reinforces that attitude: 'no regrets' is the general attitude to the passing of the old technology. Operators of equipment embodying the latest technology feel that they have a higher status, even when this is not formally confirmed by any job regrading. Working conditions are generally better, and even if they no longer practise their old skills they do not feel that they have lost them.

We had also expected to find some evidence of 'deskilling', together with resentment at the loss of traditional craft skills. But we found that new technology, while abolishing the market value of some skills, also enhances the value of others, and creates new ones. Unless one gives the word 'skill' the very narrow definition of individual dexterity, it cannot be said that there is any evidence of a long-term trend in the industrialised countries towards deskilling or towards less skilled occupations in general.

Despite the rhetoric of some union leaders, few unions in the private sector of the economy have proved to be hostile to the introduction of new technology, and some have been supportive. Any bargaining strength which the unions might have been supposed to have has been undermined by the anxiety of their members to work with new technology, thereby increasing their personal marketability. Nevertheless, trade unions are still seen by most workers, even by non-members, as a desirable form of 'insurance' in an uncertain world. In particular, they see them as a protection against arbitrary or overbearing management.

Governments have a wide range of responsibilities in modern industrialised societies, and many of these are challenged by new technology. Many of these countries, especially in Western Europe, have developed institutions and attitudes which are resistant to change. Such resistances aggravate problems such as unemployment, which governments are held responsible—at least in part—for resolving. The problems of occupational mobility which are presented by the accelerating progress of new technology provide perhaps the single most important challenge to economic policy in the advanced economies.

Governments have responsibilities for implementing new technology in their own administration, thus realising for their own citizens the enormous resource savings and improved quality of services which microprocessor technology makes possible.

In a democracy, however, they are not responsible for the

health of the body politic. That responsibility remains with the citizens themselves. It is not inevitable that new technology should lead to greater surveillance of the individual citizen by the state. Nor will new technology make more likely the introduction of centralised economic planning. But whether the citizens of a democracy will be able to resist the temptations of a society centred on machines still remains to be seen. It is early enough in the progress of new technology still to be optimistic, to believe that it can alleviate the material lot of the great majority of people, and that it will make a significant contribution to finally lifting their burdens of drudgery and poverty. It may also make possible the realisation of the traditional aspirations of human civilisation for great numbers of people. Whether this dramatic potential can, in fact, be realised will depend on the people themselves.

NOTES

1. H. L. Mencken put a similar proposition more wittily when he wrote that conscience is that still, small voice which reminds us that someone may be looking.
2. It is interesting to note that the Japanese government's contribution to the fifth-generation research programme — about $450 million spread over the ten-year period, 1981–90 — is only about one-tenth of the sum which it is estimated successive British governments spent over the nine-year period, 1975–84 in maintaining the British Leyland motor company.

References and Further Reading

ACARD (Advisory Council for Applied Research and Development), *New Opportunities in Manufacturing: The Management of Technology*, London, October 1983.

APEX (Association of Professional, Executive, Clerical and Computer Staff), *Office Technology – The Trade Union Response*, London, 1979.

(ASTMS) Association of Scientific, Technical and Managerial Staffs (ASTMS), *Technological Change and Collective Bargaining*, London, 1979.

Babbage, C., *On the Economy of Machinery and Manufactures*, Charles Knight, London, 1832.

Banking, Insurance and Finance Union, *Microtechnology: a programme for action* (undated).

Barron, I. and Curnow, R., *The Future with Micro-electronics: Forecasting the Effects of Information Technology*, London, 1979.

Boyle, G., Elliott, D. and Roy, R. (eds), *The Politics of Technology*, Longman, London, 1977.

Braun, E. and MacDonald, S., *Revolution in Miniature*, Cambridge University Press, 1978.

Braverman, H., *Labour and Monopoly Capital*, Monthly Review Press, London, 1974.

Corina, J., 'Trade Unions and Technological Change', in S. Macdonald, D. Lamberton and T. Mandeville (eds), *The Trouble with Technology*, London, 1983, pp. 178–89.

Daly, A. *et al*, 'Productivity, Machinery and Skills In A Sample Of British and German Manufacturing Plants', *National Institute Economic Review*, no. 111, February 1985, pp. 48–61.

Department Of Trade And Industry (United Kingdom), *Caring,*

Enterprise, and Winning World Market Share, London, 1985.
Drucker, P., *Innovation and Entrepreneurship*, London, 1985.

Fogarty, M., Daniel, W. and Brooks, D., *British Industrial Relations*, London, 1985.
Forester, T. (ed.), *The Information Technology Revolution*, Blackwell, Oxford, 1985.

Hall, P. (ed.), *Technology, Innovation and Economic Policy*, Philip Allan, Oxford, 1986.
Hayes, R. H. and Abernathy, W. J., 'Managing Our Way to Economic Decline', *Harvard Business Review*, July-August 1980.

Immel, A. R., 'The Automated Office: Myth versus Reality', in Forester, *op. cit.*, pp. 312–22.

Jenkins, C. and Sherman, B., *The Collapse of Work*, London, 1979.
Jewkes, J., *Government and High Technology*, Occasional Paper No. 37, Institute of Economic Affairs, London, 1972.
Jones, B., *Sleepers, Wake!: Technology and the Future of Work*, Wheatsheaf Books, Brighton, 1982.

Kanter, R. M., *The Change Masters*, Simon and Schuster, New York, 1983.

Large, P., *The Micro Revolution Revisited*, Frances Pinter, London, 1984.
Laver, M., *Computers, Communication and Society*, Oxford University Press, London, 1975.

MacDonald, S., Lamberton, D. M. and Mandeville, T. (eds), *The Trouble with Technology*, Frances Pinter, London, 1983.
Manpower Services Commission, *Adult Training in Britain*, London, 1985.
Manpower Services Commission/National Economic Development Office, *A Challenge to Complacency*, London, 1985.
Mathias, P., 'The Machine: Icon of Economic Growth', in S. MacDonald, D. Lamberton and T. Mandeville (eds), *The Trouble with Technology*, London, 1983.
Mishan, E. J., *Welfare Economics: An Assessment*, North Holland, Amsterdam, 1969.
Musson, A. E., 'Technological Change and Manpower', *History*, vol. 67, no. 220, June 1982.

National Economic Development Office (UK), *Skills and Technology: The Framework for Change*, Report issued on Behalf of Heavy Electrical Machinery, EDC, London, 1983.
National Economic Development Office (UK), *IT Futures: What current forecasting literature says about the social impact of information technology*, London, 1985.
National Economic Development Office (UK), *Advanced Manufacturing Technology*, London, 1985.
Nelson, R. and Winter, S., *An Evolutionary Theory of Economic Change*, Harvard University Press, Cambridge, Mass., 1982.
Nora, S., *The Computerisation of Society*, MIT Press, Cambridge, Mass., 1978.
Northcott, J., Fogarty, M. and Trevor, M., *Chips and Jobs: acceptance of new technology at work*, Policy Studies Institute, London, 1985.
Northcott, J., Knetsch, W. and de Lestapis, B., *Microelectronics in Industry, an International Comparison: Britain, France and Germany*, Policy Studies Institute, London, 1985.

Peters, T. J. and Waterman, R. H., *In Search of Excellence*, Harper and Row, New York, 1982.

Rajan, A., *New Technology and Employment in Insurance, Banking and Building Societies*, Gower Publishing, Aldershot, 1984.
Rothwell, S., 'Supervision and New Technology', *Department of Employment Gazette*, January 1984.

Scouller, J., *Industrial Policy in Britain*, MBA Distance Learning Course Text, pp. 7.39–7.49 on Science and Technology Policy, Strathclyde University, 1986.
Sharron, H., 'Overcoming Trade Union Resistance to Local Change', *Public Money*, London, March 1985.
Simpson, D., Love, J. H. and Walker, J., *The Effect of New Technology on Work*, Fraser of Allander Institute Research Report, Glasgow, 1986.

Trades Union Congress (TUC), *Employment and Technology*, London, 1979.

Warner, M. (ed.), *Microprocessors, Manpower & Society*, St Martin's Press, New York, 1984.
Weiner, N., *The Human Use of Human Beings*, Doubleday, New York, 1950.
Weizenbaum, J., 'The Myths of Artifical Intelligence', *New York*

Review, 27 October 1983, reprinted in Forester, *op. cit*.
Whitley, J. D. and Wilson, R. A., *Quantifying the Employment Effects of Micro-electronics*, Discussion paper 15, Institute for Employment Research, University of Warwick, 1981.

Author Index

General Index